Sedimentary Rocks and Aquifers - New Insights

Edited by Abhay Soni, Angelo Paone, Prabhat Jain and Sung-Hyo Yun

Published in London, United Kingdom

Sedimentary Rocks and Aquifers - New Insights
http://dx.doi.org/10.5772/intechopen.100920
Edited by Abhay Soni, Angelo Paone, Prabhat Jain and Sung-Hyo Yun

Contributors
Abhay Soni, Gaurav D. Chauhan, Sanjib K. Biswas, Ishaq Yusuf, Eswaran Padmanabhan, Ram Chandra Tewari, Zahid A. Khan, Prabhat Jain, Rahul Shende, Al-Mashaikie Sa'ad ZA Kader, Malebo Matlala, Angelo Paone, Sung-Hyo Yun

First published in London, United Kingdom, 2023 by IntechOpen
IntechOpen is the global imprint of INTECHOPEN LIMITED, registered in England and Wales, registration number: 11086078, 5 Princes Gate Court, London, SW7 2QJ, United Kingdom

British Library Cataloguing-in-Publication Data
A catalogue record for this book is available from the British Library

Additional hard and PDF copies can be obtained from orders@intechopen.com

Sedimentary Rocks and Aquifers - New Insights
Edited by Abhay Soni, Angelo Paone, Prabhat Jain and Sung-Hyo Yun
p. cm.
Print ISBN 978-1-80355-393-1
Online ISBN 978-1-80355-394-8
eBook (PDF) ISBN 978-1-80355-395-5

For EU product safety concerns:
IN TECH d.o.o., Prolaz Marije Krucifikse Kozulić 3, 51000 Rijeka, Croatia,
info@intechopen.com or visit our website at intechopen.com.

Meet the editors

Angelo Paone received a BS, MS, and Ph.D. from the University of Naples Federico II, Italy. He completed an isotopic specialization course at the United States Geological Survey (USGS) and a short postdoc at the American Museum of Natural History, New York, USA. He also completed a Marie Curie Fellowship at Bristol University, England, and a postdoc at the University of Naples Federico II. Dr. Paone has been a science teacher at *Liceo Scientifico Ettore Majorana Pozzuoli, Italy*, and a research professor at Pusan National University (PNU), South Korea since 2008.

Dr. Sung-Hyo Yun is a professor in the Department of Earth Science Education, Pusan National University (PNU), South Korea. He obtained a BSc from the College of Education, PNU, in 1980 and a MEd from the Graduate School, PNU, in 1982. Dr. Yun received his Ph.D. in Geology and Volcanology from the Graduate School, PNU, in 1987. Dr. Yun was the president of the Petrological Society of Korea and the Korean Society of Volcanic Hazards Mitigation.

Dr. Soni has worked in the Indian mining industry for 39 years, developing new R&D approaches and concepts. He has written three books, edited three books, and published more than 130 research papers. He has participated in more than 100 R&D projects and organized technical events for the laboratory in which he worked. He has more than 15 years of experience in scientific administration as a scientist in charge and head of research groups. Dr. Soni has chaired various official committees and was a member of the Mining Technology Group, CSIR-Central Institute of Mining and Fuel Research, Nagpur Research Centre. He is the recipient of various awards, accolades, certificates, and fellowships from several agencies. His credentials have been recognized and utilized in the industry for troubleshooting technical problems related to mines and the mining environment.

Dr. Prabhat Jain is the former regional director of the Central Ground Water Board, Ministry of Jal Shakti, Govt. of India, Central Region, Nagpur. He obtained an MTech in Applied Geology, Sagar University, India, in 1983 and a Ph.D. from Visvesvaraya National Institute of Technology (VNIT) Nagpur, India, in 2005. He has more than 34 years of experience in the field of groundwater and more than 25 years of administrative, advisory, and supervisory experience. He has made substantial contributions to the protection, preservation, and enhancement of India's water resources as a field hydrogeologist by carrying out various hydrogeological, hydrochemical, geophysical, and hydrometeorological surveys and groundwater exploration in different geological formations (aquifers). Dr. Jain has published more than 100 scientific publications, papers, and technical reports.

Contents

Preface

Sedimentary rocks are a powerful tool for better understanding the mechanism of the Earth's surface. Compared to igneous and metamorphic rocks, which are much more abundant on Earth, sedimentary rocks are more abundant on the crust. That allows their studies to unravel the processes that occur at these rocks and their relation to the tectonic in particular and plate tectonics in general. The study of the occurrence of such rocks gives us important information on their geological past and more importantly on past geological climate conditions. There are many sediments that can be used to reconstruct past climate conditions. For example, speleothems are sediments that can be used to reconstruct the past of the Quaternary climate. The study of sedimentary terrains can provide more information about recent times as well. However, there is still much to be done in terms of understanding past climate conditions and projecting those of the future.

Aquifers are an important resource for many countries. For example, in Italy, aquifers have been exploited since Roman times when aqueducts were first built. Aquifers are intimately linked to permeable sedimentary rocks (e.g., limestone) and igneous rocks (e.g., granites) like in Europe where phreatic water has been stored and cleaned by the water cycles and used by people for centuries. The unbalanced use of water, an unevenly distributed resource among continents, makes aquifers very fragile to exploit, as they are in Africa. Today, aquifers must be studied more, especially in zones where water apparently seems limited. Moreover, global warming is another negative anthropogenic climatic process that weakens aquifers, especially when temperatures are changing very quickly and drought is approaching (e.g., Spain, Greece, and Italy).

Section 1, "Sedimentary Rocks and Tectonics" includes five chapters.

Chapter 1 by Angelo Paone and Sung-Hyo Yun studies the genesis and evolution of the Pontine Plain of central Italy. The plain's morphology and sea level have persistently changed through the Pleistocene, and from the last post-glacial period, the coast started to retreat. The most recent hydrogeological maps (1957, 1977, 2007) clearly show that water is saturating the plain and this will cause serious mobility problems in the future. The chapter highlights several hazards of the plain, including volcanic, seismic, coastal erosion, and sinkhole hazards, as well as water contamination hazards (e.g., arsenic and fluorine), which have increased the prevalence of cancers in the region. Thus, there is an urgent need for monitoring in the area (i.e., Latina Province).

Chapter 2 by Kader explains a new and abnormal type of sandstone in the world derived from the Zagros Ophiolite Thrust Belt in Northeast Iraq, which extends from the Himalayas in Pakistan passing through Iran, Iraq, Turkey, and along Europe. This sandstone reflects the composition of the ophiolite sequence composed essentially of fresh olivine-pyroxene and ultrabasic igneous rock fragments. Ophiolite sandstone composed mainly of serpentinite rock fragments (about 40%) was reported in

Italy only. The chapter deals with the origin and mechanism of the preservation of the high percentages of the fresh olivine-pyroxene and ultrabasic igneous fragments grains.

Chapter 3 by Zahid A. Khan and Ram Chandra Tewari studies the primary sedimentary features (cross strata and planar strata) used in estimating the hydraulic geometry parameters of the Barakar River, including channel form, channel dimensions, channel paleoslope, flow velocity, boundary shear stress, and friction factor. In the chapter, the hydraulic geometry parameters of the Barakar River are computed quantitatively and the trunk of the river is three-dimensionally modeled. The chapter shows that the river channel that carried sediments in the basin had a sinuosity of 1.361 in the south–southwest part and progressively became more sinuous in the northern and northeastern parts. The estimated catchment area of the Barakar River lies between 4452.69 and 14,749.83 km^2 and paleo-discharge between 4510 and 22,070 m^3/sec with a maximum of 66,000 m^3/sec and a mean annual flood of 183,700 m^3/sec. Consequently, the estimated flood plain area of the Barakar River is about 66,000 km^2. The paleo-morphological and paleo-hydrological data suggests that the Barakar River is remarkably broad, extending far and wide with a northward drainage flow along the northeastern border of India (Sikkim, Assam, and Bangladesh) towards the Tethys Ocean.

Chapter 4 by Ishaq Yusuf and Eswaran Padmanabhan discusses the goal of rock fabric characterization. It describes the spatial and geometric distribution of pore attributes and their impact on rock petrophysical parameters such as porosity, permeability, and water saturation. The goal of rock fabric characterization is to describe the spatial and geometric distribution of pore attributes as they impact petrophysical parameter variations. The chapter shows that multiple fabrics occur in a single lithofacie in the form of a fabric domain. It also infers that these fabric domain types are responsible for the multiple occurrences of hydraulic fluid units (HFUs). No long-range process exists that aligns the microscopic internal fabric or micro fabric architecture among grain aggregates in sedimentary rock.

Chapter 5 by Sanjib K. Biswas and Gaurav D. Chauhan presents the tectonic framework of the Indian Plate since the breakup of Gondwanaland in the Late Triassic. Its evolution continued between its separation from the African Plate in the Early Cretaceous and its collision with the Eurasian Plate on the north in the Late–Middle Eocene. Subduction and plate motion are responsible for the activation of primordial faults depending on the related stress field. The major tectonic zones (TZs) are the Himalayan TZ, Assam-Arakan TZ, Baluchistan-Karakoram TZ, Andaman-Nicobar TZ, and Stable Continental Region (SCR) earthquake zone. The orogenic belt circumscribing the northern margin of the Indian Plate is highly tectonised as the subduction of the plate continues due to northerly push from the Carlsberg Ridge in the southwest and slab-pull towards northeast and east along the orogenic and island arc fronts in the northeast. Three major fault zones from north to south are the North Kathiawar fault - Great Boundary fault (along the Aravalli belt) zone, South Saurashtra fault (extension of Narmada fault), Sonata-Dauki-Naga fault zone, and Tellichery-Cauvery-Eastern Ghat-T3-Hail Hakalula-Naga thrust zone. The neotectonic movements along these faults, their relative motion, and displacement are the architects of the present geomorphic pattern and shape of the Indian craton. With the reactivation of this shear zone, the two proto-cratonic blocks are subjected to relative movement as the plate rotates anticlockwise.

Section 2, "Aquifers," includes two chapters.

Chapter 6 by Abhay Soni et al., studies areas of the basaltic aquifer through aquifer mapping. Areas under study dominated by igneous rock consisting of basaltic lava flows have low storage potential. The groundwater storage and movement within the aquifer can be classified according to their hydraulic condition, which varies from the geological formations (lithological condition). This chapter discusses the management of basaltic aquifers where the groundwater storage potential is very low compared to other aquifers using aquifer mapping.

Chapter 7 by Malebo Matlala analyzes the groundwater dynamics in aquifers of Southern Africa. Groundwater resources are indispensable not only in water-scarce or water-stressed countries but also globally as dependable reservoirs and an alternative resource of freshwater. This chapter assesses the spatiotemporal variability of groundwater resources within two of the biggest transboundary aquifers that South Africa shares with its neighboring countries. The Karoo-Sedimentary Transboundary Aquifer (KSTA) and the Stampriet Transboundary Aquifer System (STAS) were studied over a period of 72 years from 1948 to 2020. The results show that the use of remote sensing (RS) techniques coupled with geographic information system (GIS) applications are invaluable where there is a dearth of scientific data and information. Furthermore, they can be used for the monitoring, management, and protection of groundwater resources on local, regional, and international scales.

Dr. Angelo Paone
Research Professor,
Pusan National University,
Busan, Republic of Korea

Dr. Abhay Soni
Chief Scientist,
Nagpur Research Centre,
CSIR-Central Institute of Mining and Fuel Research (CSIR-CIMFR),
Nagpur, Maharashtra, India

Dr. Prabhat Jain
Central Ground Water Board,
Central Region, Ministry of Jal Shakti,
Government of India,
Nagpur, Maharashtra, India

Dr. Sung-Hyo Yun
Professor,
Pusan National University,
Busan, Republic of Korea

Section 1

Sedimentary Rocks and Tectonics

Chapter 1

Introduction to the Sedimentary Geology, Genesis, and Evolution of the Pontine Plain in Relation to the Alban Hills Volcanism and Associated Hazards

Angelo Paone and Sung-Hyo Yun

Abstract

The Pontine Plain was reclaimed by Fascists in the 1930s. The Plain morphology and sea level have persistently changed through the Pleistocene, and from the last post-glacial period, the coast started to retreat. The most recent hydrogeological maps (1957, 1977, and 2007) clearly show that water is saturating the plain, and this will cause serious mobility problems in the Pontine Plain in the future. Several hazards of the plain are listed in this chapter. These include volcanic, seismic, coastal erosion, and sinkhole hazards, as well as water contamination hazards (e.g., As F), which have increased the number of cancers. There is an urgent need for monitoring in the area (i.e., Latina).

Keywords: pontine plain, sediments, Alban Hills, paleoclimatology, hydrogeology, hazards

1. Introduction

The term Pontine Plain generally refers to the flat territory of Lazio between Terracina, Anzio, Circeo, and the Lepini Mountains obtained through reclamation of the Pontine Marshes in the 1930s. The history began more than two thousand years ago with the ancient people of the Volsci and ended in the first decades of the twentieth century with the "integral reclamation" of the 1930s (**Figure 1**) [1, 2]. As shown in **Figure 2**, the northward motion of the Adriatic Plate is affected by the interaction with the African Plate, which pushes from the south, and also by that of the Aegean Microplate, which moves toward the southwest at high speed [3]. The tectonics of Italy has shaped its morphology, especially the Apennines and Alps chains. Conversely, the Italian Plains (e.g., Po Plain and Pontine Plain) formed during the recent period of Pleistocene climate variation. The Roman area of interest lies between the Apennine chain and the Tyrrhenian Sea. The Apennine chain is a complex structural unit, with a series of thrusts toward the E–NE that formed mainly between the upper Miocene and lower Pliocene. Following the tectonic shortening

Figure 1.
Recent geological map of the study area 1:1000000). From project CARG Piana Pontina. North arrow like in ***Figure 3****.*

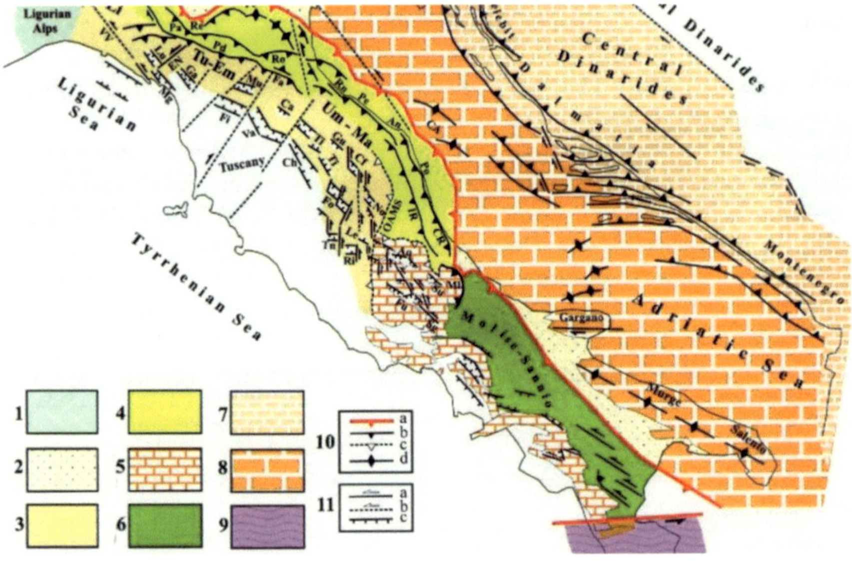

Figure 2.
Main structural and tectonic features of the central Apennines and the peri-Adriatic zones: (1) Ligurian and southern Alps; (2) Foredeep basins; (3, 4) axial and outer belts of the northern Apennines; (5) Latium–Abruzzi and southern Apennines carbonate platforms; (6) outer belt of the southern Apennines; (7) Dinarides carbonate platforms; (8) Adriatic foreland; (9) Calabrian arc; (10) compressional features: A = outer front of the Alps and Apennines; b, c = active and inactive thrusts, d = fold axes; (11) Transcurrent and extensional features: A, b = active and presumably active strike-slip faults; c = normal faults. Scale 1:1000000 (modified from [3, 4]. North arrow like ***Figure 3****.*

phases, the internal sector of the chain underwent a progressive process of extension toward the west with the formation of the Tyrrhenian back-arc basin. In particular, the Lazio anti-Apennine south of the Tiber is composed of the anti-Apennine volcanic zone with the Alban Hills, an isolated group that retains traces of volcanic activity including various solfataras. Southeast of the Alban Hills, the topography rises into the Lepini Mountains, and further south into the Ausoni and Aurunci Mountains [3].

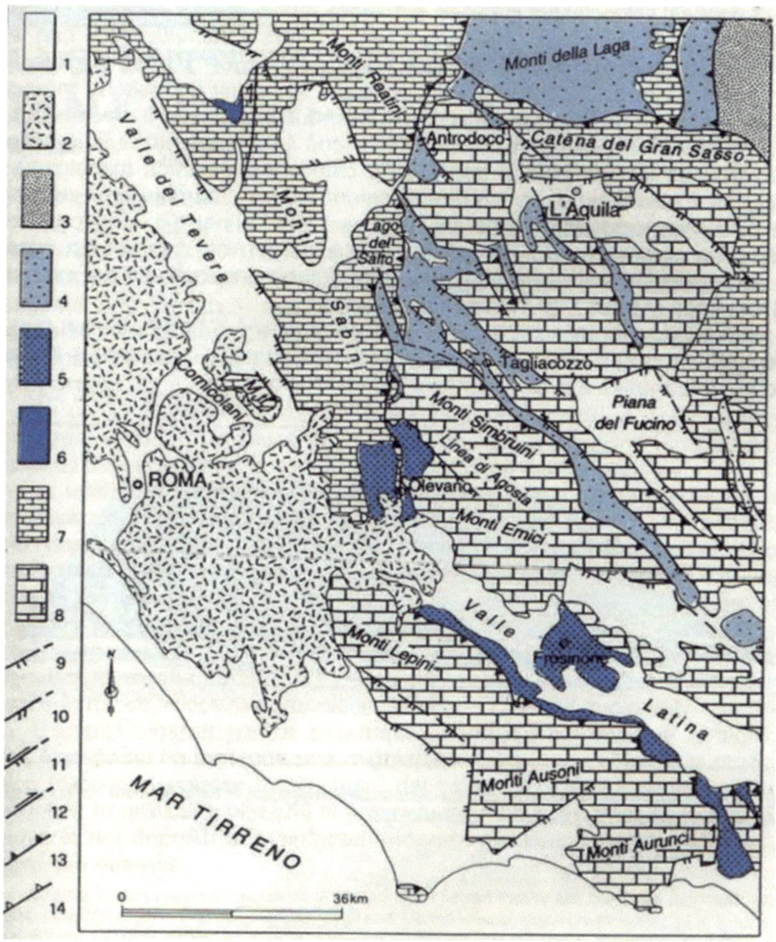

Figure 3.
Geological-structural scheme of Central Italy. 1-Plio–Pleistocene continental marine deposits and recent alluvial coverings; 2-vulcanites (Pleistocene); 3-syntectonic terrigenous deposits (Cellino formation, lower Pliocene); syntectonic terrigenous deposits (clay–arenaceous formation, upper Tortonian p.p.-upper Messinian); syntectonic terrigenous deposits (Frosinone formation, upper Tortonian p.p); 6-syntectonic terrigenous deposits (Marly–arenaceous formation, Burdigaliano p.p-Langhiano); 7-stratigraphic succession in transition facies (upper Triassic–lower Miocene); 8-stratigraphic succession in carbonate platform facies (upper Triassic–middle Miocene); 9-direct fault; 10-transtensive fault; 11-fault with complex kinematics; 12-transcurrent fault; 13-overrun; 14-perception. Modified after Ref. [4].

The Pontine Plain extends between the Lepini Mountains and the sea. Beyond this, the promontory of Mount Circeo forms a short (5 km long) limestone chain. **Figure 3** shows the sedimentary terrains surrounding the study area (Pontine Plain). We will focus on the origin of the Pontine Plain in relation to the sedimentology and paleoclimatology, along with a volcanological outline.

2. Geological history of the pontine plain and Alban Hills volcano

The topics of interest are the Pontine Plain and Alban Hills, which had a much younger genesis. Pliocene–Quaternary, than the sedimentary terrains outcrop-ping around the Pontine Plain (**Figure 4**). At the upper Pliocene–Pleistocene, the shoreline

Holocene	
Pleistocene	Upper
	Middle
	Lower
Pliocene	Gelasian
	Piacenzian
	Zanclean

Figure 4.
Chronologic interval covering the formation of the pontine plain (Pliocene–Pleistocene). Modified from (https://www.isprambiente.gov.it/scala_tempi_geologici?current_era=Miocene).

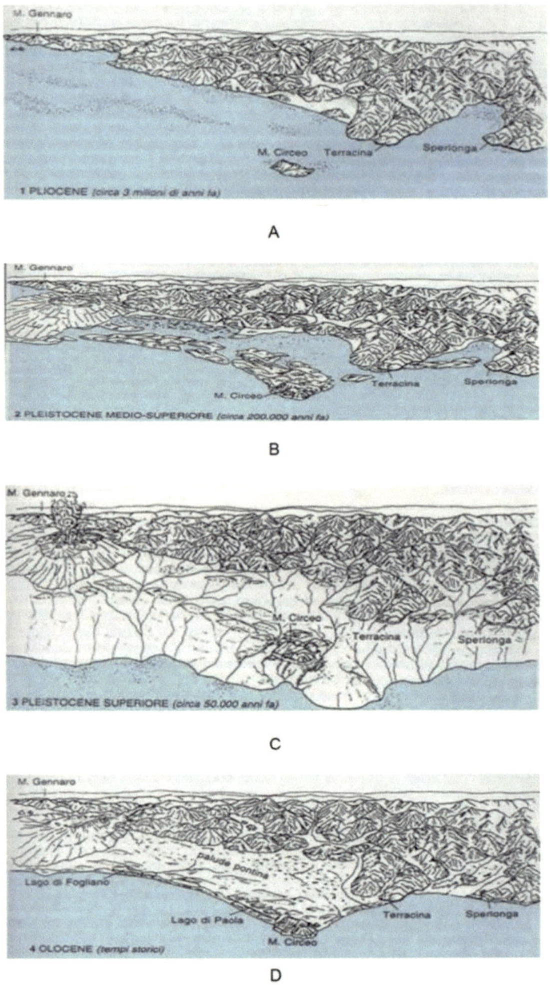

Figure 5.
A geological model of the pontine plain around the Pliocene (3 ma ago). B geological model around the middle Pleistocene (200,000 years ago). C geological model around the end of the Pleistocene (50,000 years ago). D geological model around the Holocene (10,000 years ago). Modified from https://www.parcocirceo.it/pdf/Piano%20del%20Parco%203%20pag%2029-57.pdf.

had reached the Lepini and Ausoni carbonate reliefs, which were still in the uplift phase. Mount Circeo was a limestone island separated from the mainland by a large area of sea (**Figure 5A**). The sediment contributions of the drainages flowing into the sea and the contribution of clastic materials linked to the initiation of activity of the Alban Hills volcano caused the first partial filling of the marine basin and the formation of a series of coastal bars. The volcanic activity of the Alban Hills began shortly before the beginning of the Pleistocene (**Figure 5B**). The lowering of sea level associated with the onset of the cold climate phases of the Pleistocene caused the shoreline to advance approximately 10 km. The entire area was transformed into a marshy area with some water courses that reached the sea and formed incised valleys. The volcanic activity at Alban Hills continued (**Figure 5C**). With the marine transgression that began 10,000 years ago at the end of the Würminaglacial stage, the shoreline regressed, assuming an articulated trend by occupying the previous river valleys. The deposition of new coastal strips led to the formation of a new, different, and more rectilinear shoreline with the isolation of the Fogliano, Monaci, Caprolace, and Paola Lakes (**Figure 5D**, **Table 1**). Dunes separated the lakes from the sea, as can be seen in the case of Circeo (**Figure 6A**). A geomorphologic model for Pontine Lake (Sabaudia, **Figure 6B**) was constructed. **Figure** 7 shows digital reconstructions of the Pontine Plain with coastal lakes formed during the regression of the coastline during the Holocene. Moreover, the warm period brought erosion of the coastline, which is still occurring.

Era	10^3 years	Period	Climate Period	Occurrences
Quaternary	10	Holocene	Post-glacial	Glacial deposits, recent fans at the mouth of the streams, moraines at the mouth of the valleys
	100	Upper Pleistocene	Wurm	Last ice age Completely decalcified stratigraphic deposit
	180		Inter-Glacial Riss-Wurm	
	250		Riss	
	400 450 550 600	Middle Pleistocene	Inter Glacial Mindel-Riss Mindel Inter Glacial Mindel-Gunz Gunz	Stratigraphic deposit originated by the rocks present in place Conoids at the mouth of valleys to form terraces on the high plain
C			Inter Glacial	Glacial deposits First ice age of Pleistocene
			Donau-Gunz	
	750	Lower Pleistocene	Donau	
	800			
		Upper Pliocene		

Table 1.
Ice age succession with the sedimentary terranes formed.

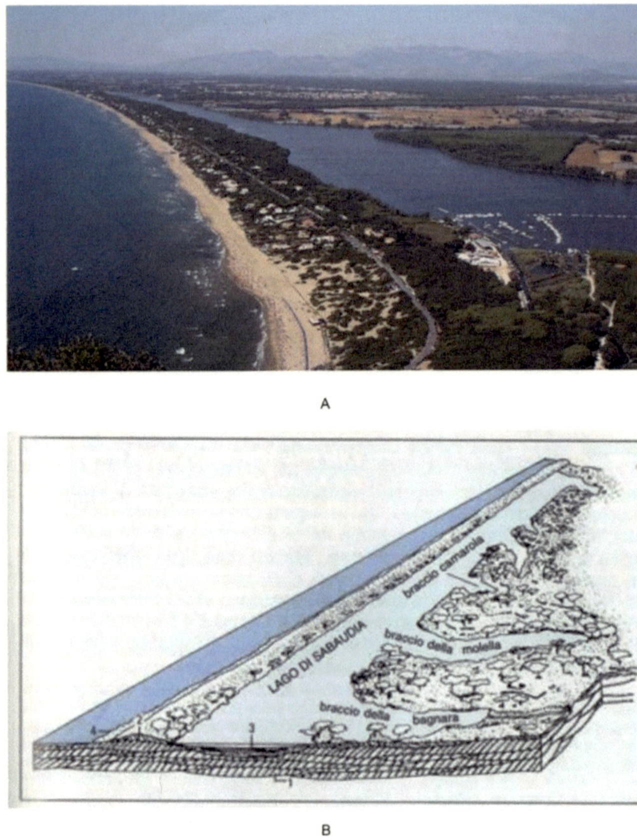

Figure 6.
A recent coastal dunes. B Sabaudia Lake model. 1-sandy deposits; 2-sandbar of recent dune; 3-current lake silt–sand deposits; 4-current beach. Modified from the http://www.zaininspalla.it/articoli.php?url=articoli/geologia/circeo.htm&titolo=Note%20geologiche%20e%20geomorfologiche%20dell%E2%80%99area%20pontina&PHPSESSID=frpledcgn761q5jl2ljspo9gf1.

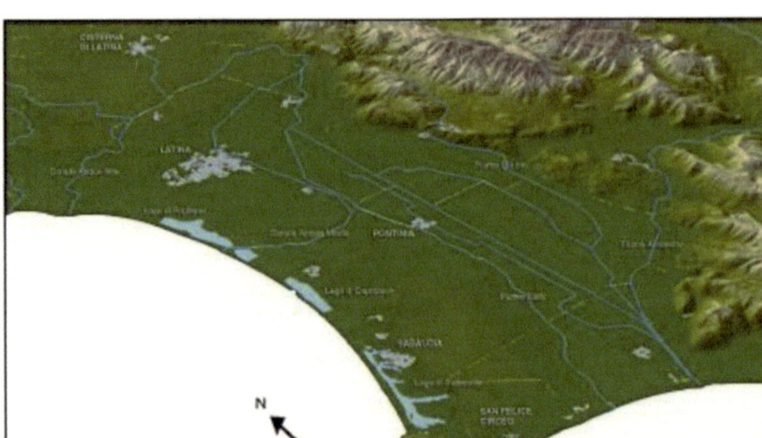

Figure 7.
Digital elevation model of the pontine plain highlighting the coastal lakes. Modified from [5].

3. Hydrogeology of the pontine plain

The carbonate outcrops of the Monte Lepini complex were lowered several 100 m below sea level by a system of vertical faults that accommodated dynamic extension. The depression, thus, formed was filled by clayey sediments dating to the Pliocene and Calabrian (gray-blue clays) during the marine ingress and regression phases. The Quaternary formations of the coastal area consist mostly of sandy continental formations (Upper Pleistocene) deposited by wind in the form of large dunes on preexisting coastal–brackish sandy deposits consisting of organic limestone sands (Middle–Lower Pleistocene). The entire sandy complex, which varies in thickness from a few tens of meters to about a hundred meters, is called the "Duna Antica." Proceeding inland, the recent formations of the continental environment are replaced by deposits formed in fluvial–marsh environments. These consist of alternations of sandy and sandy–clayey strata along with travertine formations. In the marshy environment, formed in the foothill depressions, numerous peaty levels formed and contribute to triggering its marked subsidence [6, 7]. Some additional pertinent information is listed below:

A. Toward the north, the Acque Alte channel forms a watershed bordering the volcanic aquifer system of the Alban Hills.

B. Two different Quaternary aquifer systems, separated by the Sisto River, which constitute a watershed with two different hydrogeological basins, can be distinguished in the Pontine Plain (Duna Antica and Pedemontano) [8, 9].

C. Below the Quaternary aquifers, groundwater is confined to the basal aquifer of lower limestones of the Lepini Mountains system, which is fed partially by lateral infiltration and reverse percolation of the alluvial aquifer multilayer in the foothills area.

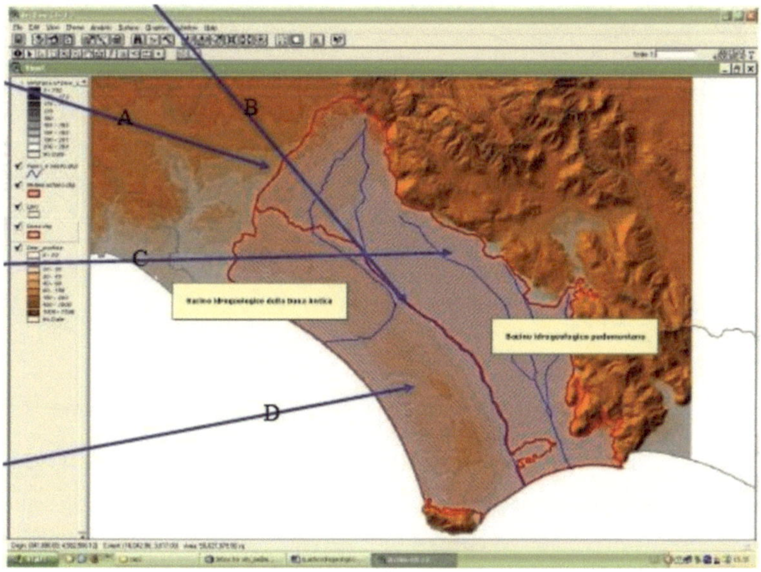

Figure 8.
ESRI GIS hydrogeological map modified from [5] permitted to show it.

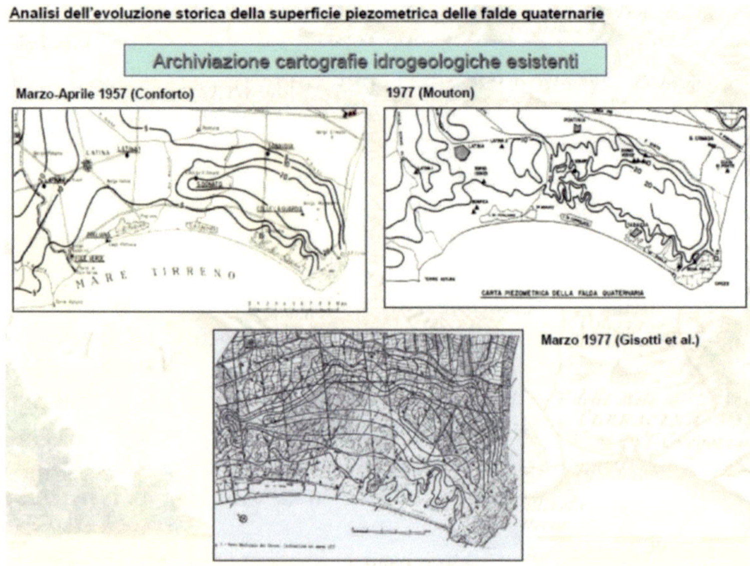

Figure 9.
Historic evolution of the piezometric surface within the pontine plain from 1957 to 1977 (Conforto, 1957; Mouton, 1977, Gisotti et al., 1977) [9, 12].

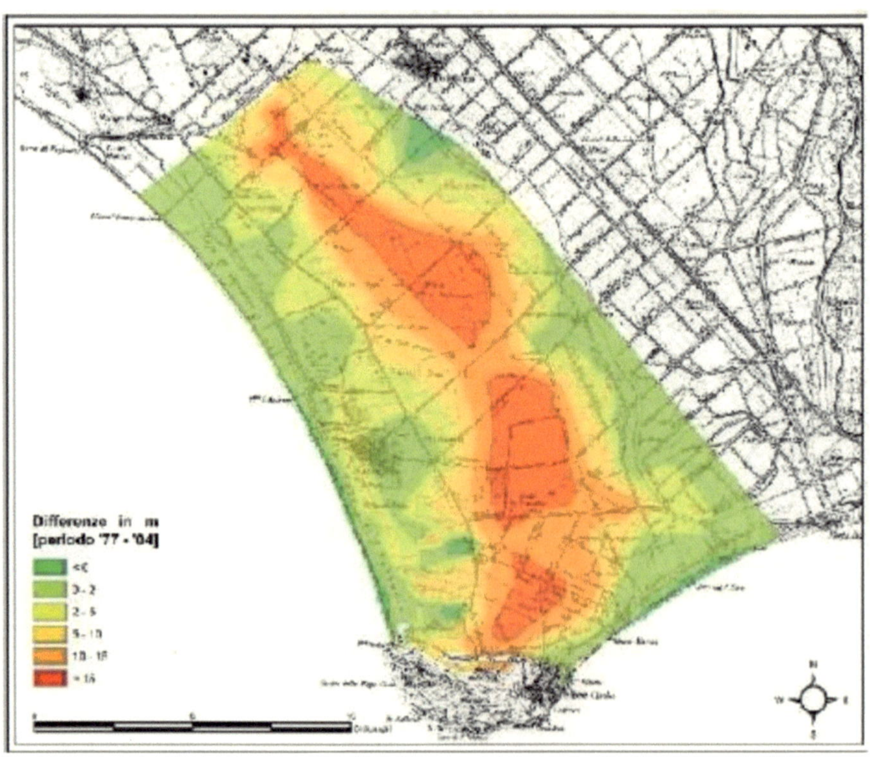

Figure 10.
Comparison of results of hydrogeological studies between 1977 and 2007 [5].

D. The aquifer of the Duna Antica is considered a hydrogeological basin filled solely by direct rainfall (**Figure 8**) [8, 9]. Broadly, the aquifer of the Pontine Plain evolved considerably from the time of the first hydrogeology studies (Conforto, 1957 [5, 9, 10] to 1977) (Mouton, 1977; Gisotti et al. 1977 [5, 9–11]) (**Figure 9**). **Figure 10** shows a comparison of the hydrogeology between 1957 and 1977. The total error of the comparison process can be assumed to have a maximum value of about 6 m, although the computed error value is usually much lower than 6 m. Given the apparent phenomenon of lowering saturation levels, management decisions are needed.

4. Volcanological context of the pontine plain region and volcanic hazards

The most important volcanic threat is the quiescent Alban Hills volcanic complex (**Figures 11** and **12A, B**). New results for the Alban Hills volcano (Roma, Italy), which

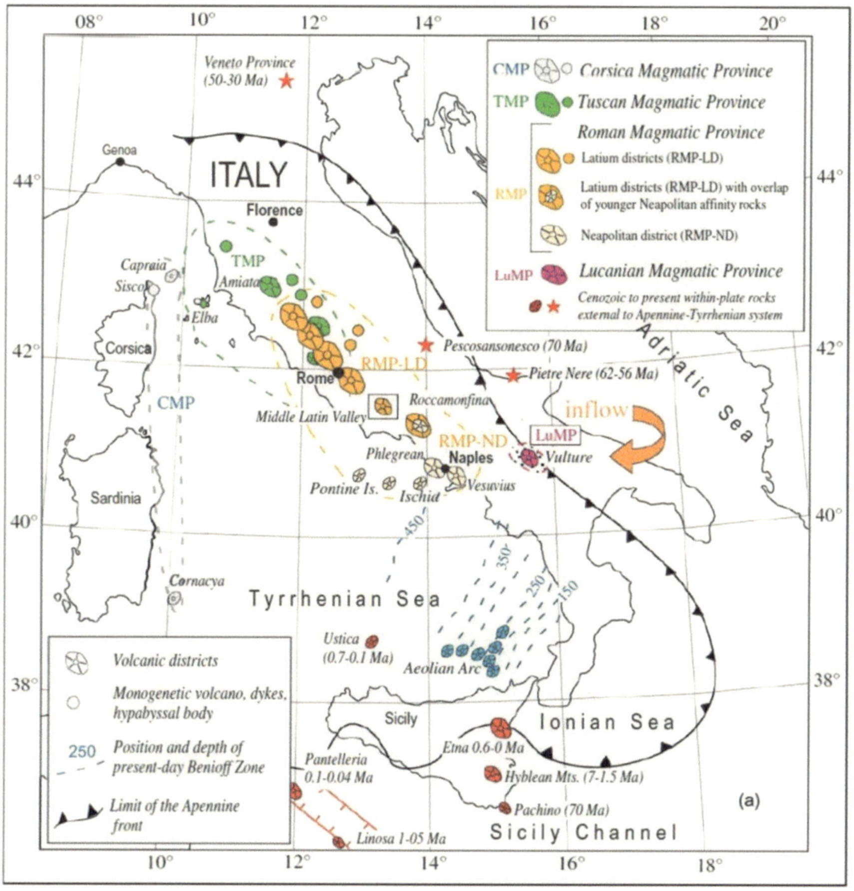

Figure 11.
Volcanic map of Italy. Distribution of the Plio-Pleistocene volcanic rocks in the Italian peninsula and geological sketch map of the Sacco River valley (i.e., Alban Hills and middle Latin Valley volcanic field associated with the pontine plain). Drawn after Refs. [13, 14]. Main monogenetic volcanoes are represented by cinder cones, tuff rings, and small plateau-like lava flows [15].

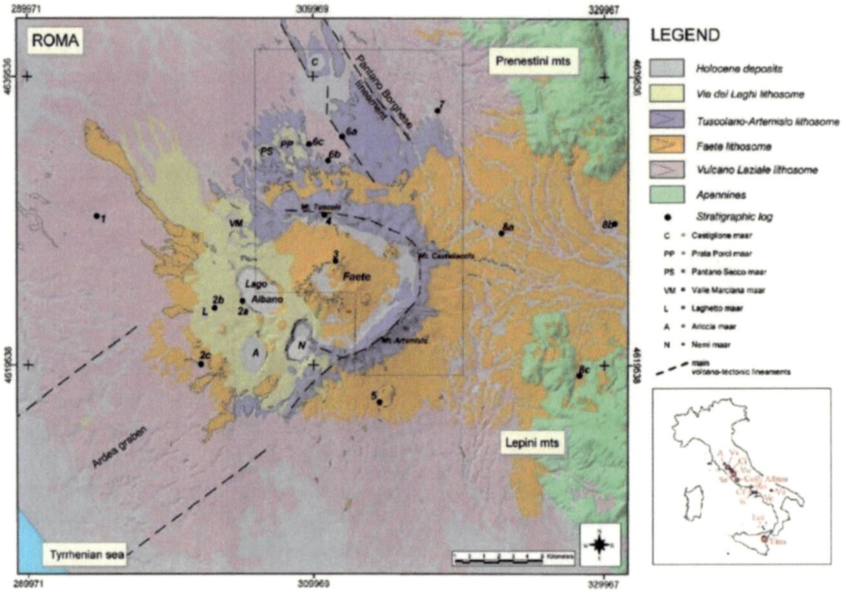

A

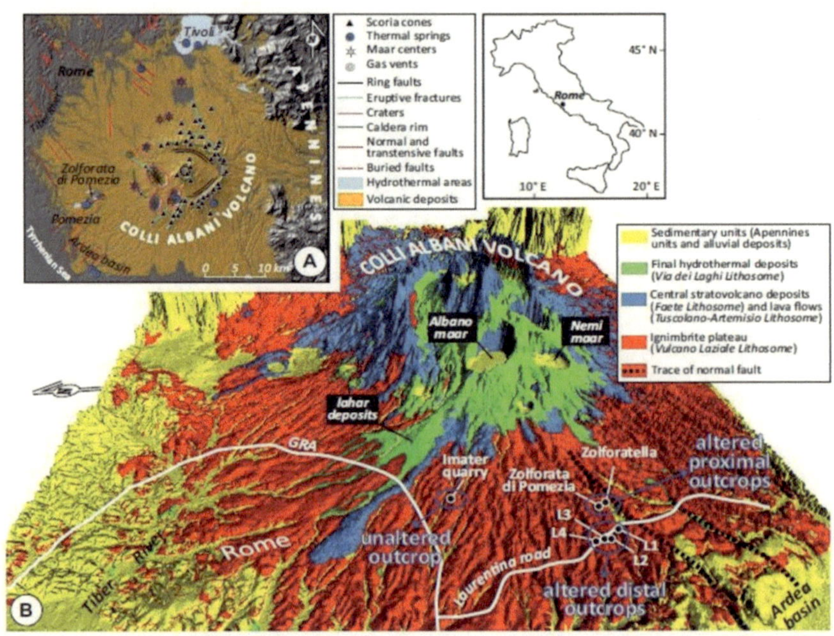

A

Figure 12.

A geological setting of the Colli Albani volcano Central Italy. B three-dimensional view of the Colli Albani volcano (10 times vertical exaggeration). The study sites are located in the west-southwestern sector of the volcano, namely, the Imater quarry, Zolforata di Pomezia mine, Zolforatella, and Laurentina 1–4 outcrops (L1, L2, L3, and L4). Except the Imater quarry (unaltered term), hydrothermal alteration and sulfate or sulfide mineralization affect all other study sites. Alteration occurs at Zolforata di Pomezia and Zolforatella (proximal outcrops), whereas L1 to L4 are distal outcrops. GRA is the main highway encircling the city of Rome [16, 17].

was surveyed for the 1:50,000 scale Geological Map of Italy (CARG Project) [18], provide insights into the caldera evolution. Colli Albani, a quiescent volcano, was active at approximately 600 ka. The eruptive products are consistently mafic (<50% SiO_2); nevertheless, the morphology and dominant explosive eruptive style are like those of felsic calderas [19]. The volcano is a composite, containing multiple superposed edifices or lithosomes. The oldest edifice (Volcano Laziale (VL), ca. 600–350 ka) is a 1600 km^2 plateau of low-aspect ignimbrites (VEI 5–7) with a central caldera. After the last large eruption (>50 km of deposits), forming the Villa Senni Eruption Unit ignimbrites at ca. 355 ka, two edifices were built within the caldera. The first is the horseshoe-shaped Tuscolano–Artemisio (TA) composite edifice (or lithosome), which

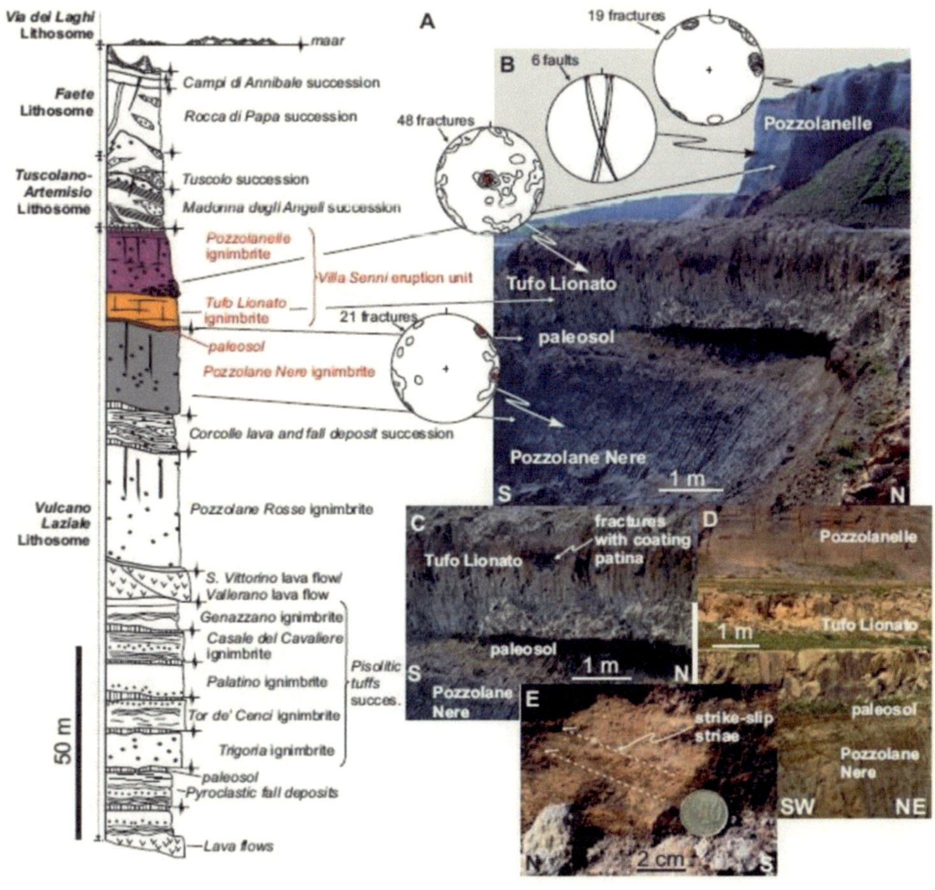

Figure 13.
A composite stratigraphic column of the Colli Albani volcano (simplified after Giordano et al., 2006). The studied section (from Pozzolane Nere to Pozzolanelle) is emphasized with colors. B photograph of the studied section within the Imater quarry. Diagrams with contours are Schmidt polar nets (lower hemisphere) representing the density (red: Maximum data density) of extensional fracture poles measured in the studied volcanic units. The diagram with the cyclographic data is a Schmidt net (lower hemisphere) showing the attitude of subvertical N-striking strike-slip faults observed in the quarry exposure. Red dots along the cyclographic data are fault subhorizontal striae. C–D close-up photographs showing details from the exposure shown in B. In C, note the presence of fractures, particularly in the Tufo Lionato ignimbrite. E close-up photograph showing a N-striking fault surface with subhorizontal striae [16, 17].

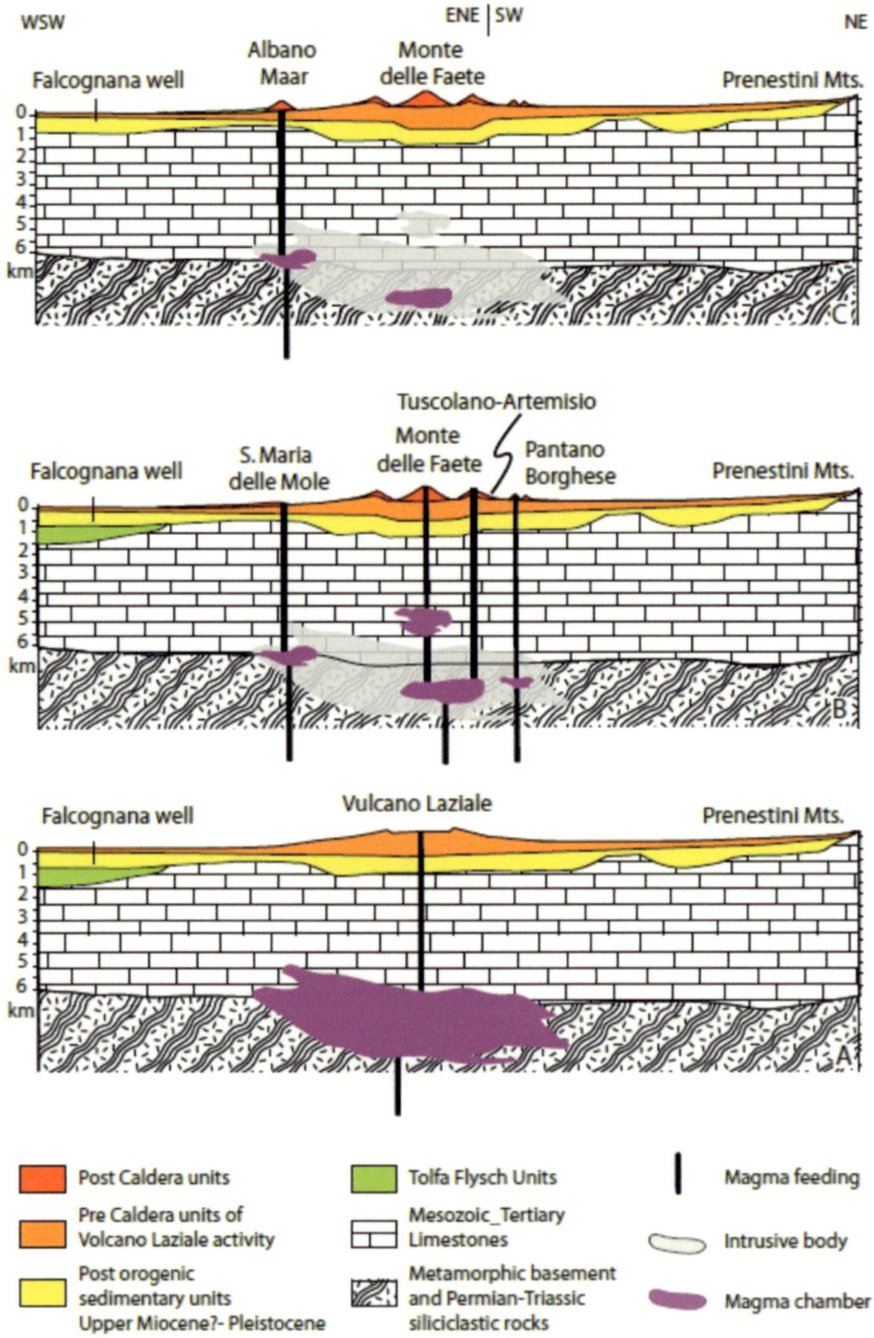

Figure 14.
Cartoon showing the possible volcanological and magmatological evolution of the Colli Albani plumbing system. Here, the magma chamber and plumbing systems have been reconstructed with petrological, isotopic, and geochemical data and geological and geophysical interpretations of the geological and structural setting of the Alban Hills prevolcanic units. Xenoliths from the Alban Hills volcanic rocks along with information from deep wells drilled in the area allowed the reconstruction of the stratigraphy of the substratum of the volcano, whereas the thicknesses of the main geological units were derived from gravimetric and seismic data [14].

consists of coalescing, peri-caldera, fissure-related scoria cones interbedded with lava flows; the fissure system forms two segments controlled by regional faults. The second is the steep-sided Faete stratovolcano (949 m a.s.l.), which filled the caldera. The TA and Faete lithosomes, which partly interfinger, were emplaced at ~350–260 ka. Their products indicate reduced eruption rates relative to the VL period and a change to effusive and mildly explosive eruptions. The most recent and still active phase of phreatomagmatic activity formed overlapping maars and tuff cones along the western and northern slopes of the volcano, collectively named the *via* dei Laghi composite lithosome [19]. The Alban Hills caldera is a polyphase caldera: (1) A piecemeal caldera is associated with large-volume ignimbrites of the VL edifice, and the present shape of the caldera is related to the Villa Senni eruption. (2) The TA composite edifice, erupted from peripheral caldera fissures, is unrelated to explosive phases of caldera collapse. The TA final products cover a morphologically stable caldera wall. The peripheral fractures feeding the TA composite edifice are interpreted as volcano-tectonic structures activated during the late-stage downsagging of the caldera. Reduced eruption rates during the TA and the Faete stages (1–10 km^3/ka compared to >100 km^3/ka for the VL edifice) suggest that diminished recharge of the magma chamber may have induced prolonged deflation and downsagging of the caldera floor and the opening

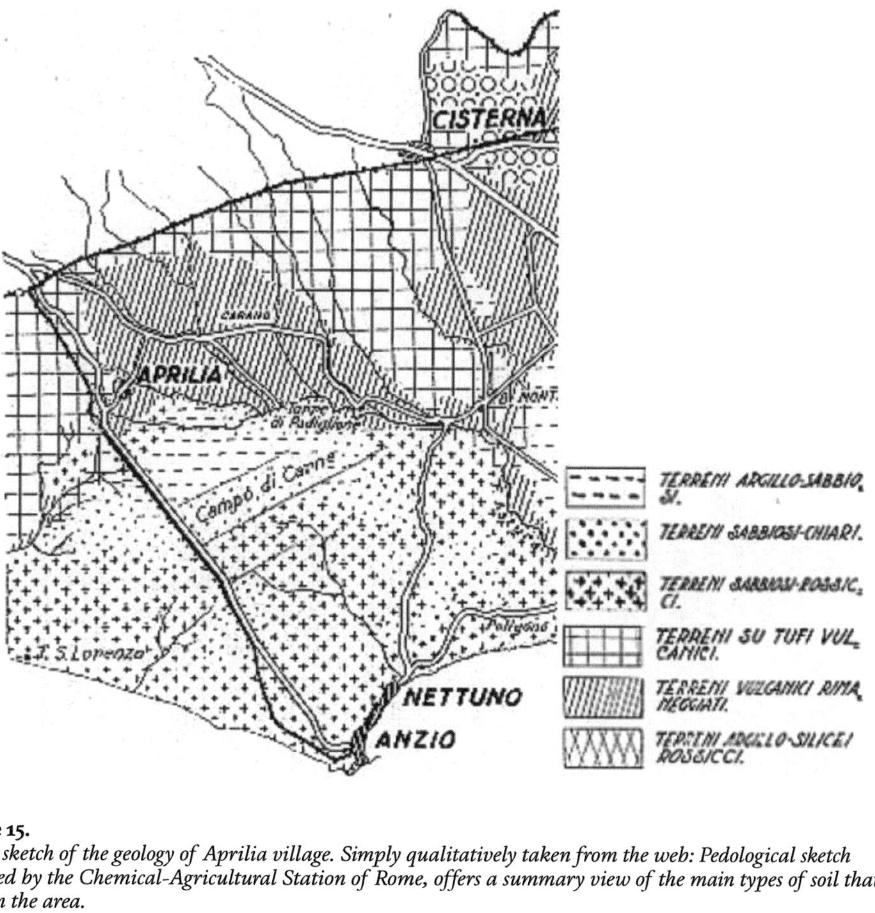

Figure 15.
Simple sketch of the geology of Aprilia village. Simply qualitatively taken from the web: Pedological sketch compiled by the Chemical-Agricultural Station of Rome, offers a summary view of the main types of soil that occur in the area.

of outward-dipping peripheral fractures [19]. Based on this interpretation, the TA edifice represents the surface expression of ring dikes at depth. The absence of similar fissure structures along the western caldera rim may be related to the deep geometry of the inward-dipping ring faults in those areas, which therefore were not favorably oriented for magma intrusion during a period of general subsidence. "In contrast, the subsequent and still active phreatomagmatic phase, which emplaced the *via* dei Laghi composite edifice and is active on the western side of the caldera, may produce resurgent conditions that could affect Aprilia village and other areas near the volcano" (**Figure 13**) [16]. A volcano model of Alban Hills [12] suggests that such a volcano could potentially affect all of the territory around it, a densely inhabited area, in addition to Aprilia village (Latium Province) (**Figures 13–15**).

5. Hazards of the pontine plain

5.1 Seismic hazards

The Earthquake Hazard Map of Lazio produced by the Istituto Nazionale di Geofisica e Vulcanologia (INGV) indicates that the Pontine Plain has a degree of danger that is among the lowest in Lazio (0.050–0.125). Although these data are reassuring, we cannot forget the threat of earthquakes from the Alban Hills, a volcanic system that is currently dormant but that could resume activity at any time (**Figure 16**).

5.2 Volcano hazards

Given the state of the Alban Hills volcanic system and the presence of volcanic materials in Aprilia (the volcanic soils of Aprilia are remodeled, but some in-place material remains), there is clear evidence that an explosive eruption such as those produced by the Alban Hills could easily reach the city of Aprilia. The presence of products associated with the occurrence of peripheral volcanic activity in numerous perforations (pyroclastic rocks and travertines arranged on numerous horizons) causes us to hypothesize the regional existence of evolved volcanism capable of influencing the geochemical characteristics of the waters of the Pontine slopes.

5.3 Coastal erosion

The last 10,000 years (Holocene) have been a warm period, and the climate warming process has been accelerated by global warming attributable to the release of CO_2 into the atmosphere through the industrial period, which in turn resulted in some retreat of the coastline. Coastal erosion has recently occurred in some parts of the Pontine Plain, so the Latina coastline has been nourished by the Latina Municipality. There is strong evidence that the coast is retreating in the Latium coastal region, so nourishing is urgently needed along this coast.

5.4 Sinkholes

Catastrophic collapse features, called sinkholes in the international literature, are a type of instability that affects the surface of the soil as sudden collapses occur over a short period of time (6–24 hours). These differ from the normal karst forms because of their suddenness and conditions of initiation and development. Sinkholes

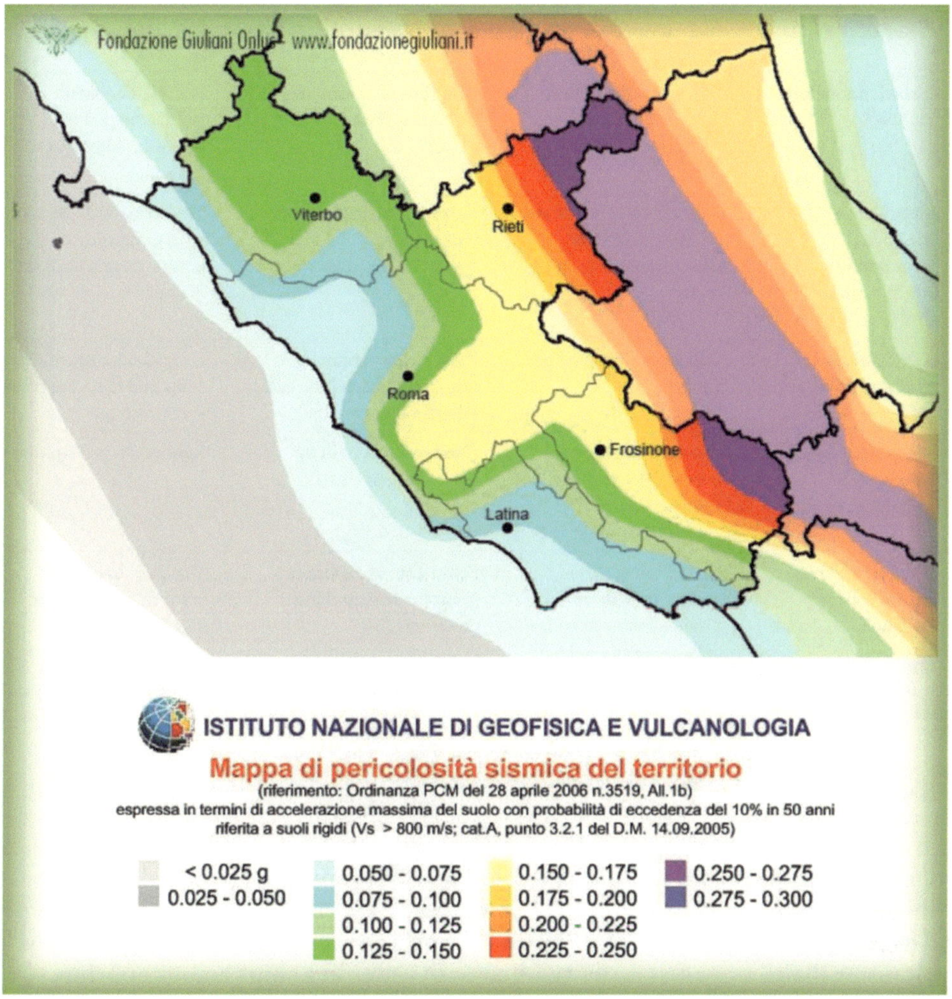

Figure 16.
Lazio seismic hazard map (from INGV https://www.google.com/search?q=Mappa+della+pericolosit%C3%A0+sismica+del+Lazio&tbm=isch&chips=q:mappa+della+pericolosit%C3%A0+sismica+del+lazio,online_chips:zonazione+sismica).

can be classified according to their genesis, their interaction with the topographical surface, and the hydrogeological structure of the area in which they form. Their impact on the landscape, and even more on the populations in the areas most at risk, is very high, as demonstrated by the numerous bibliographies available, which often describe huge economic losses and, in some cases, losses of human life, although cases of this type have not yet occurred in Italy. If we also consider the fact that sinkholes are neither random nor extemporaneous phenomena and that their localization, morphometry, and evolutionary typology, as well as their temporal recurrence, clearly demonstrate that they occur in specific geological conditions (sinkhole-prone areas) and in the presence of various geological factors (sinkhole-triggering factors), it is clear that, even in areas such as Italy, it is necessary to reach a profound understanding of the development and triggering modalities these phenomena, even more profoundly if we consider that many of the sinkholes that occur in the susceptible areas of Italy have some peculiar characteristics that substantially differentiate them

Definition	Main genetic processes	Sedimentological setting	Morphology
Solution sinkhole/ doline	Chemical dissolution of karst bedrock	Outcrops of carbonate/ evaporitic rocks	Enclosed depression funnel-shaped with a flat bottom. Thin cover of red soil
Subsidence sinkhole, cover-subsidence sinkhole, cover sinkhole	Subsurficial dissolution or collapse of karst voids in the underlying rocks	Karst carbonatic bedrock with non-consolidated cover (i.e., sand, gravel).	Topographic depression with dimension of some tens of meters both in diameter and depth
Collapse/cave-collapse sinkhole, collapse doline	Collapse of the roof of caves	Karst cave overlain by lithoids deposits (i.e., tuff, limestone, travertine)	Deep sinkholes with steep walls and truncated cone vertical shape
Rock-subsidence sinkhole, subsidence doline	Collapse on cohesive, permeable and non-soluble rocks placed over soluble sediments	Karst cave overlain by cohesive deposits (i.e., clay, silt)	Funnel-shaped sinkholes
Piping sinkhole, cover-collapse sinkhole	Upwelling of water and gases with a piping process followed by the collapse of the soil	Sediments of different granulometry and less cohesive	Variable shape from smooth depression to steep sinkholes

Terms used to describe the sinkholes, the main genetic processes, the sedimentological setting and the morphology (partly taken from Gunn, 2004).

Table 2.
Terminology of sinkhole.

from those studied in the rest of the world [20]. The terminology, which is vast, was schematically classified in **Table 2** [21]. **Figures 17** and **18** illustrate the diverse types and evolutions of sinkholes [20]. **Figure 19** shows the typical sinkhole typology of the Pontine Plain. Doganella's cavity filled with water in a short amount of time. However, the sinkhole of Doganella di Ninfa did not change very much during the few months after its formation. The Doganella chasm appeared in December 2003 [21]. Additional details on all sinkholes in the Piana Pontina are available in several sources (**Figure 20**) [22–24].

5.5 Drinking water quality

Water analyses from the Agenzia Regionale per la Protezione dell'Ambiente del Lazio (ARPA) broadly provided data (shown in **Table 3**) of the turbidity, microbiology, arsenic content, and fluoride content of the drinking water of villages within the Latina province including Aprilia, Sermoneta, Latina, Cisterna di Latina, and others. These waters are extremely hazardous. As shown by these data, the relevant institutions must take major control of these waters, especially given the concern about contaminants that would be highly dangerous if consumed in drinking water. The hazards of these waters have been confirmed by the Azienda Sanitaria Locale (ASL). In fact, the contamination revealed by ARPA and ASL indicates that the sampling locations are associated with high rates of cancers. The results indicate a need to conduct more in-depth epidemiological studies (resident studies) by initiating a monitoring campaign to estimate the levels of arsenic and fluorine in the resident

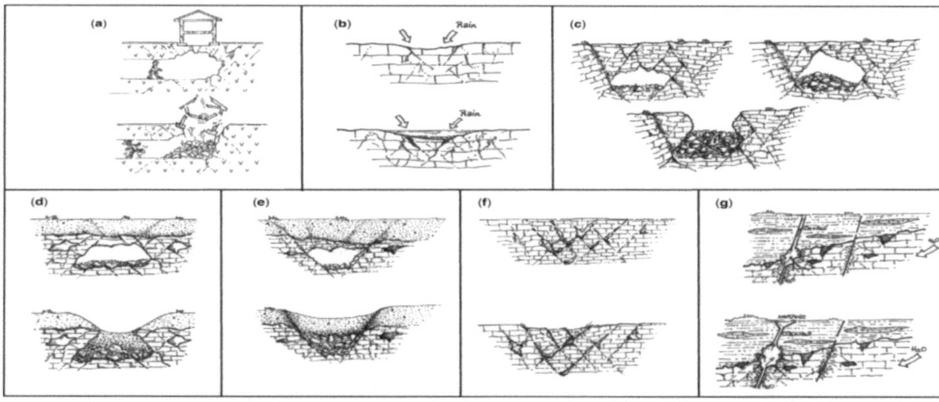

Figure 17.
Classification of sinkholes based on the literature and field observations of Italian case studies. (a) Anthropogenic sinkhole; (b) solution doline/sinkhole; (c) cave collapse; (d) cover collapse; (e) cover subsidence; (f) rock subsidence; and (g) piping sinkhole [20].

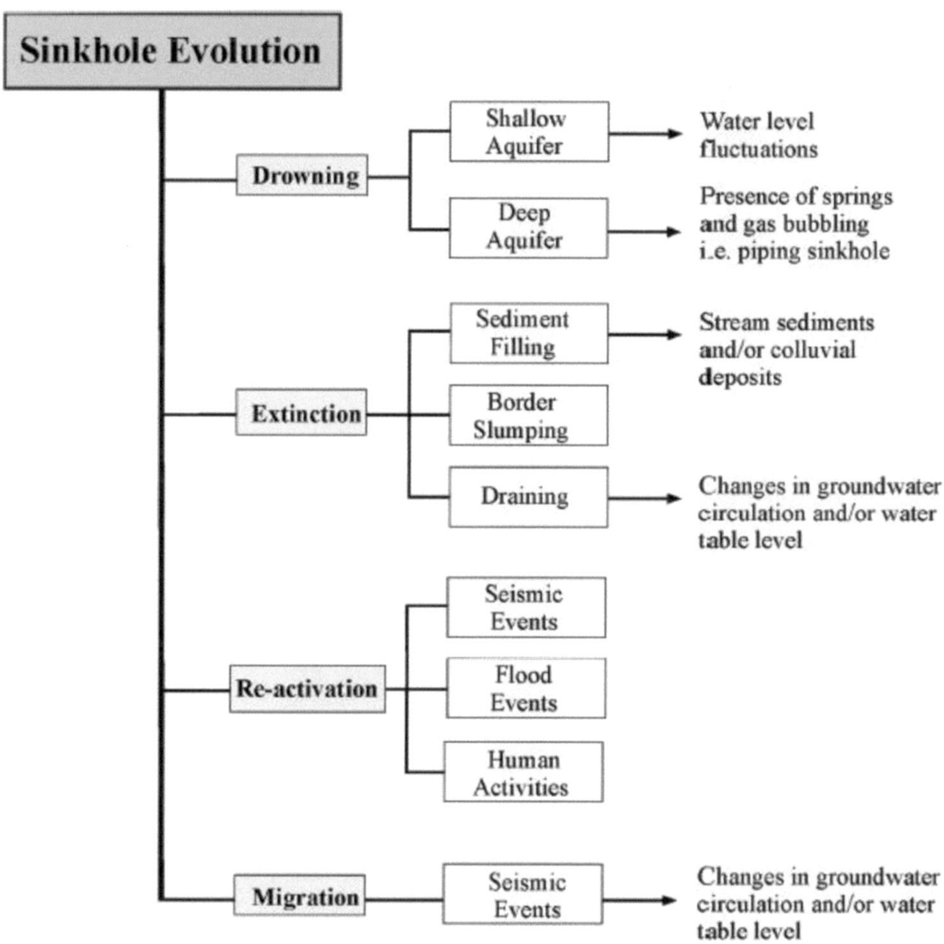

Figure 18.
Schematic of sinkhole evolution. Modified from [20].

Figure 19.
Typical sinkhole of the pontine plain. Dimension 3x3 m.

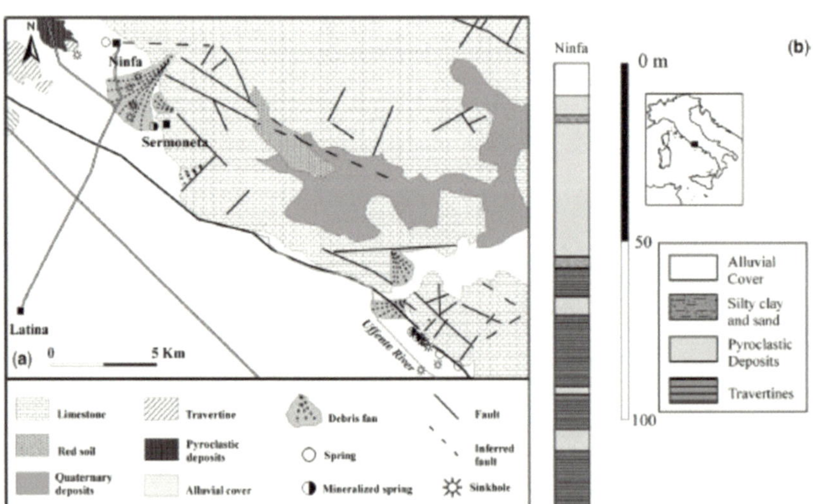

Figure 20.
Areas in the pontine plain at risk of catastrophic sinking (based on history): (1) Doganella–Via Ninfina (Comuni di Sermoneta e cisterna di Latina). (2) Contrada Ciocco–Contrada Talci, San Michele. (3) Gricilli–Mezzaluna–Tenuta Isabella–Cotarda.

Comune	Nr. di campioni non conformi triennio 2007–2009 *(D.lg. 31/01 e modifiche O.Lgs. 27/2002)*			
	Torbidità	**Contaminazione microbiologica**	**Arsenico**	**Floruri**
Aprilia	0	1	11	6
Cisterna	0	14	114	1
Latina	1	16	0	0
Ser moneta*	0	7	5	0

Fonte: Banca dati ARPA latina - (*) Il dato di Sermoneta si riferisce al triennio 2008–2010.

Table 3.
Concentrations of harmful elements in the drinking water of Latina villages.

population. Survey results indicate plausible health effects in the resident populations of the municipalities of Rome and the Latium provinces, which have high arsenic levels. The effects refer to chronic exposures through the past decades, during which arsenic levels could have been higher than the current levels. The results indicate the need for continuous monitoring of water arsenic levels and public health interventions to ensure compliance with the limits established by the legislation currently in force (directive 98/83/EC, As <10 μg/L).

6. Conclusion

The Pontine Plain is a fragile and young territory, and its fragilities, unrestrained anthropogenic use, and natural changes have brought various types of disasters, as described in this chapter. The fascist reclamation of the Pontine Plain dramatically affected the demography of the area, considering that the people from North Italy (Lombardia, Veneto, and Emilia-Romagna regions) who moved to the plain would have preserved the area. Before this time, the area was abandoned and attended by criminals. Currently, although the area is well populated and tended, global climate change (e.g., [25]) and the geological circumstances continue to modify the environment, bringing several hazards (seismic activity, volcanic activity, coastal erosion, sinkholes, and drinking water contamination) attributable to both nature and recent human activities. Most importantly, the plain could again experience a submerged period, which began in the post-glacial period (Holocene) and is being strengthened by global warming.

Acknowledgements

Chris Hawkesworth is thanked for some comments for the style of the paper.

Author details

Angelo Paone* and Sung-Hyo Yun
Pusan National University, Republic of Korea

*Address all correspondence to: angelo.paone1@gmail.com

References

[1] Pennacchi A. Canale Mussolini. Milano: Arnoldo Mondadori Editore; 2010. p. 460

[2] Pennacchi A. Parte seconda. In: Canale Mussolini. Arnoldo Mondadori Editore; 2015. p. 425

[3] Viti M, Mantovani E, Babbucci D, Tamburelli C. Quaternary geodynamics and deformation pattern in the Southern Apennines: Implications for seismic activity. Bollettino della Società Geologica Italiana. 2006;**125**:273-291

[4] Cipollari P, Cosentino D. Miocene unconformities in the central Apennines: Geodynamic significance and sedimentary basin evolution. Tectonophysics. 1995, 1995;**252**(1-4):375-389

[5] Sappa G, Rossi M, Marcolini L. GIS application to evaluate the irrigation water demand in Pontina plain by multisource data elaboration. L'Acqua. 2007;**3**(2007):33-42

[6] Sappa G, Rossi M, Coviello MT. Effetti ambientali del sovrasfruttamento degli acquiferi della Pianura Pontina (Lazio). In: Aquifer Vulnerability and Risk, 2nd International Workshop. 4th Congress on the Protection and Management of Groundwater. Reggia di Colorno - Parma, 21-23 Settembre 2005. 2005

[7] Brunamonte F, Serangeli S. Evoluzione naturale ed intervento antropico nello sviluppo dei fenomeni di subsidenza nella Pianura Pontina. Mem. Soc. Geol. It. 1996;**51**:823-836

[8] Alimonti C, Perotto C, Gazzetti C, Marinucci E. Captazioni e risorsa idrica nel bacino di Mazzocchio. In: Progetto Monitoraggio Acque Superficiali Interne e Costiere della Provincia di Latina. Roma: Gangemi editore spa; 2007. p. 129

[9] Sappa G, Rossi M. Carta idrogeologica del sistema acquifero della Piana Pontina. 100.000. Regione Lazio: Università della Sapienza di Roma; 2007

[10] Sappa G, Rossi M. Idrogeologia del Sistema Acquifero Della Piana Pontina "Hydrogeology of the Pontina Plain Aquifer System". Rome, Italy: Dip. di Idraulica, trasporti e strade. Sapienza Università di Roma – Regione Lazio, Assessorato all'Ambiente e alla Cooperazione tra i popoli. Direzione Regionale Ambiente; 2007

[11] Mounton J. Contributo allo studio delle acque sotterranee dell'Agro Pontino e Romano. In: Atti Convegno: "L'acqua per la Pianura Pontina: situazione e prospettive". Consorzio Bonifica: Latina; 1977. pp. 115-156

[12] Boari E, Avanzinelli R, Melluso L, Giordano G, Mattei M, De Benedetti AA, et al. Isotope geochemistry (Sr-Nd-Pb) and petrogenesis of leucite-bearing volcanic rocks from "Colli Albani" volcano, Roman Magmatic Province, Central Italy: Inferences on volcano evolution and magma genesis. Bulletin of Volcanology. 2009;**71**(9):977-1005. DOI: 10.1007/s00445-009-0278-6

[13] Pasquare G, Serri G, Vezzoli L. Carta geologica dell'area della MediaValle Latina. Scala 1:50 000. Progetto finalizzato geodinamica. Sottoprogetto: sorveglianza dei vulcani e rischio vulcanico. In: Carte tematiche sul vulcanismo recente. Milan: Istituto di Geologia, Universita' di Milano; 1985

[14] Boari E, Conticelli S. Mineralogy and petrology of Mg-rich calc-alkalic, potassic, and ultrapotassic associated rocks: The middle Latin Valley monogenetic volcanoes, Roman

Magmatic Province, Southern Italy. Canadian Mineralogist. 2007;**45**: 1443-1469

[15] Peccerillo A. Plio-Quaternary Volcanism in Italy: Petrology, Geochemistry, Geodynamics. Berlin: Springer; 2005. p. 365

[16] Giordano G, De Benedetti AA, Diana A, Diano G, Gaudioso F, Marasco F, et al. The Colli Albani mafic caldera (Roma, Italy): Stratigraphy, structure and petrology. Journal of Volcanology and Geothermal Research. 2006;**155**:49-80

[17] Vignaroli G, Aldega L, Balsamo F, Billi A, De Benedetti AA, Filippis LD, et al. A way to hydrothermal paroxysm, Colli Albani volcano, Italy. Geological Society of America Bulletin. 2014;**127**(5-6):672-687, published online on 6 November 2014. DOI: 10.1130/B31139.1

[18] Funiciello R, Giordano G. Note illustrative della Carta Geologica Italia alla scala 1:50.000. Foglio 374 Roma. Rome, Italy: Università Roma Tre. Dipartimento di Scienze Geologiche; 2008. p. 160

[19] Funiciello R, Giordano G, editors. The Colli Albani Volcano. London: Geological Society; 2010. Special Publications of IAVCEI 3. 400 pages plus geological map

[20] Parise M, Gunn J, editors. Natural and Anthropogenic Hazards in Karst Areas: Recognition, Analysis and Mitigation. Vol. 279. London, Special Publications: Geological Society; 2007. pp. 23-45. DOI: 10.1144/SP279.4 0305-8719/07/$15.000

[21] Gunn J. Contributory area definition for groundwater source protection and hazard mitigation in carbonate aquifers. In: Parise M, Gunn J, editors. Natural and Anthropogenic Hazards in Karst Areas: Recognition, Analysis and Mitigation. Vol. 279. London: Geological Society, Special Publications; 2007. pp. 97-109

[22] Nisio S. I fenomeni di sprofondamento: stato delle conoscenze ed alcuni esempi in Italia centrale. Il Quaternario. 2003;**16**:121-132

[23] Nisio S, Caramanna G, Ciotoli G. Sinkholes in Italy: First results on the inventory and analysis. Italian Journal of Geosciences. 2007;**125**(3):273-291

[24] Nisio S, Salvati R. Fenomeni di sprofondamento catastrofico. Proposta di classificazione applicata alla casistica italiana. In: Proceeding: State of the Art on the Study of Sinkhole. Rome; 2004

[25] Stocker TF, Qin D, Plattner G-K, Alexander LV, Allen SK, Bindoff NL, et al. Technical summary. In: Stocker TF, Qin D, Plattner G-K, Tignor M, Allen SK, Boschung J, Nauels A, Xia Y, Bex V, Midgley PM, editors. Climate Change 2013: The Physical Science Basis. Contribution of Working Group I to the Fifth Assessment Report of the Intergovernmental Panel on Climate Change. Cambridge, United Kingdom and New York, NY, USA: Cambridge University Press; 2013

Chapter 2

Abnormal Ophiolite (Olivine/Pyroxene Rich) Sandstone NE Iraq: An Approach to the Origin and Tectonosedimentary Evolution of Zagros Foreland Basin

Al-Mashaikie Sa'ad ZA Kader

Abstract

Unusual Paleocene ophiolite sandstone rich in olivine/pyroxene identified in Zagros Thrust Belt (NZTB) in NE Iraq. NZTB is regionally extended from Iran to Alpen Belt. Kolosh sandstones are controlled by progressive thrusting during late Cretaceous-Paleocene. Zagros thrust sheets composed of ophiolites, oceanic crust, basaltic flows, and ash sequences. Kolosh sandstones reveal high percentages of fresh olivine-pyroxene grains accompanied by igneous intrusive and volcanic ultrabasic-basic fragments, which are reported for the first time in NE Iraq and along ZTB. Olivine, pyroxene, ultrabasic igneous altered, serpentine and chlorite fragments, heavy minerals (includes chrome spinal), anorthite, and labradorite all together composed about 70% of the mineralogical composition. Sanidine, anorthoclase, quartz and cristobalite, argillaceous, carbonate and chert fragments all together composed (12.25%), supported by argillaceous matrix (16.53%), which are derived from mantle and oceanic crust/ophiolite sequences from NE Iraq, emplaced during late Cretaceous with arc volcanism, which subjected to rapid submarine erosion and deposition. Intense wave action accelerated the erosion of beach rocks, and concentrate the heavy minerals insitue that slumped to deeper margins. Identified lithofacies types, grouped in four associations, slope/submarine channel, inner, outer fan, and hemipelagic/pelagic, respectively, represented progressive upward transgression from slope to basin plain systems controlled by progressive thrusting.

Keywords: Kolosh, Badelyan, pyroxene/olivine-rich sandstone, Zagros thrust, ophiolites

1. Introduction

The Kolosh formation is studied here for the first time in the Badelyan area in the NE extremities of the high-folded zone, very near to the suture of the NZTB (**Figure 1**) [1], which is widely distributed in NE Iraq. It consists of successive sandstones and mudstones interbeds [2]. The previous workers suggested no signs of turbidites and graded beddings in the formation and classified them as molasses sediment deposited in a narrow platform basin [3–5].

Ophiolite sandstones are very rare in the world, which was reported in Italy (Val Marecchia Nappe) composing high percentages of serpentine and ultrabasic fragments [6]. The ophiolite sequence is considered as the main source of intrusive ultrabasic/basic and basic volcanic grains are basically derived from mantle, oceanic crust, and from volcanic arcs. The green sand is composed of relatively high percentages of olivine grains that were recently recorded only in the Hawaii Islands, where the lava flows are composed of olivine basalt type [7]. High active storm and submarine waves made rapid erosion and deposition of the heavy fractions, for example, olivine grains in the shallow environment.

This chapter discusses sandstone units of the Kolosh formation, which contain high percentages of fresh olivine, pyroxene, and ultrabasic/basic igneous grains as an unusual case in Iraq and in Neotethys extension. It tries to interpret the potential

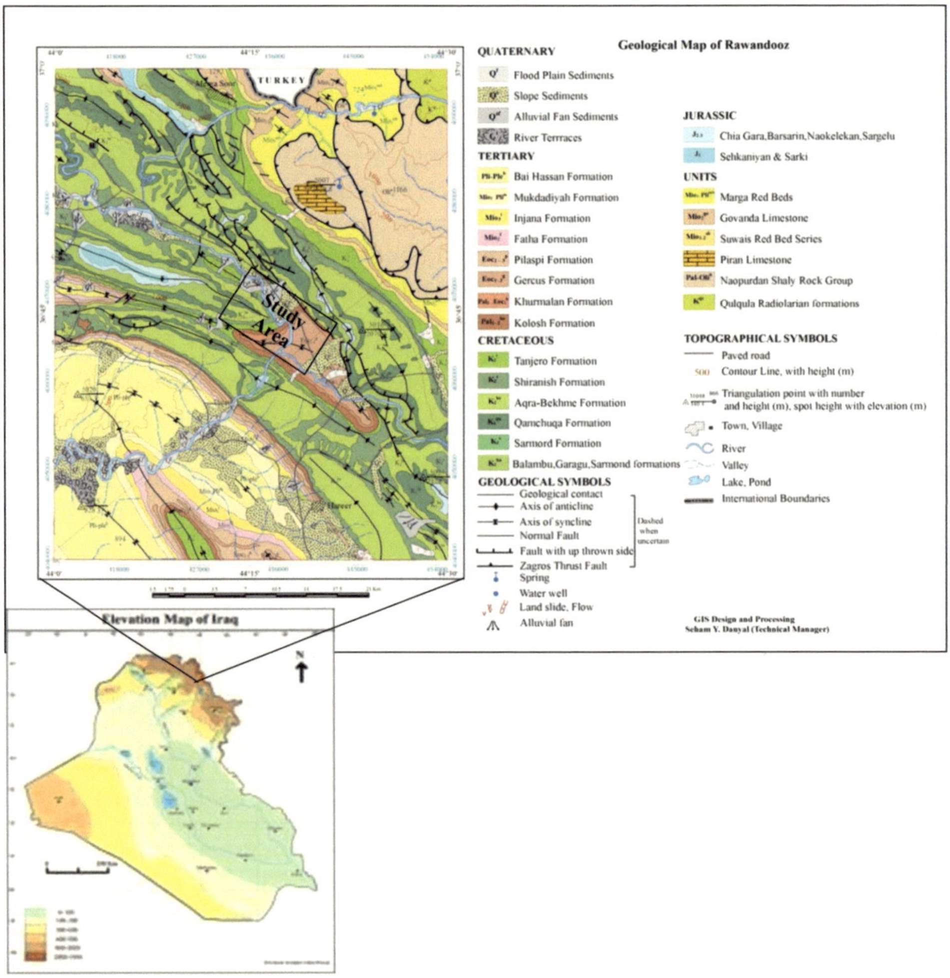

Figure 1.
The geological map of Dyana-Rawandoz-Sidakan-Rayat areas in the suture of the Zagros thrust belt NE Iraq shows the study area (rectangular) (Geol. map of Erbil and Mahabad GEOSURV, compiled by Sissakian et al. [1].

source rocks of the olivine and pyroxene concentration related to tectonic thrusting and the mode and mechanism of deposition. The paper studied the facies and associations relative to tectonosedimentary evolution in response to thrusting tectonism and seismicity. Arrangement of the facies sequences of the Kolosh formation in the studied area refers to deep marine turbidites rejecting the previous suggestion of no turbidities and no graded beddings. The turbidites facies is supported by characteristic sedimentary structures of turbidity origin as discussed later.

2. Stratigraphic review

Previous studies described the Kolosh formation at the type section in the Kolosh village in Koi-Sanjaq area, NE Iraq. [8] were firstly described as the Paleocene-Lower Eocene deposits in the high-folded zone. Buday [4] said that the formation was composed of 400 m thick blue shales and green sandstones. The above upper 174 m is of Sinjar limestone formation underlying with 140 m thick of intertonguing Sinjar and Kolosh formations. In Derbendikhan area, a section of 1000 m thick was described by Ref. [9], including many conglomerate beds with mudstones, siltstones, argillaceous, and detrital limestone interbeds. The Kolosh formation is encountered in several oil fields of northern Iraq [3, 4]. The formation was deposited in a relatively narrow rapid sinking basin trending NW-SE and was superimposed on platform margins and separated from the geosyncline by uplifted emerged and eroded lines [3, 4, 9]. No turbidite and graded bedding signs were suggested by Ref. [4], which diminished to an open sea calcareous sequence.

Previous paleontological studies ascertained the middle Paleocene age [10]. According to Ref. [11], the lower contact in the type locality is unconformable and transgressive. However, the authors [4] ascertained conformable upper contact with Paleocene-Lower Eocene limestone. The firstly discovery of olivine/pyroxene-rich sandstones of the Kolosh formation in the Badelyan locality was carried out by the study of Ref. [2].

3. Materials and methods

This study is based on the stratigraphic and petrographic analysis of Paleocene sedimentary sequences of the well-exposed section in the Badelyan area NE Iraq (**Figure 2**). A total of 170 m thicknesses were studied and logged, which is located directly beneath the suture of the Zagros thrust belt. The field works cover the measurement of bed thicknesses and the total thickness of section. Moreover, litho and facies types, sedimentary structures, sedimentary sequences, and bed extensions are identified and measured. The field works of the measured section, lithological description, thickness variations, and stratigraphy of the sequences are studied and summarized in **Figure 2**. Descriptions of field features accompanied by the results of microscopic studies have been used for facies analysis and paleoenvironmental construction. For subsequent studies, a total of 50 hand specimens were collected. Sampling intervals were generally between 1 and 3 m. Additional data, including macroscopic rock descriptions and ichonological data, have been studied as well as trace fossils in the field work.

Petrographic facies analyses were carried out in 45 thin sections and reported all mineralogical constituents, matrix, and cement types. Detailed photomicrographs

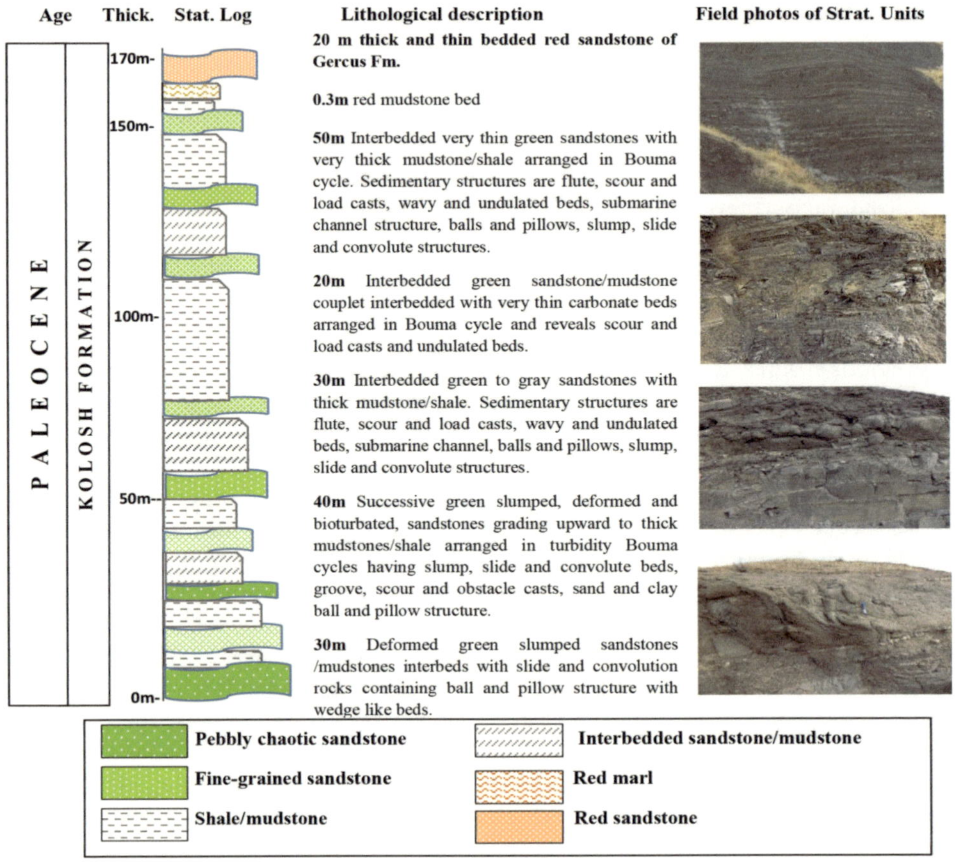

Figure 2.
The stratigraphic section shows different stratigraphic units of the Kolosh formation in Badelyan area, NE Iraq.

are picked up for all mineralogical details. Abbreviations for names of rock-forming minerals used to point to the types of minerals in the photomicrographs are followed [12]. The sandstone classification of Ref. [13] as well as the provenance and tectonic setting of Ref. [14] cannot be applied because of the high content cf. 25% of olivine and pyroxene grains.

4. New stratigraphic details

Location: the studied section lies 5 km N of Badelyan village, 150 km from Erbil Governorate, NE Iraq.

Thickness of the Kolosh formation in Badelyan section is about 170 m (**Figure 2**).

Lithology: the formation consists of mudstones/shales, muddy sandstones, pebbly sandstones, chaotic mudstone, fossiliferous sandstone, and thin pelagic carbonate beds. Very thin siliceous horizons are observed within the mudstone beds. All beds of shales and sandstones are arranged in graded turbidity cycles. Most beds are deformed, revealing characteristic sedimentary structures of turbidite origin.

Sedimentary structures are identified based on Refs. [15–20] as main characteristic of turbidity origin cf. load and groove casts, ball and pillows, graded beddings, slump slide and convolute beds, scour, submarine channels, and wavy and undulated beds. Most of the sandstone and mudstone horizons are graded upwards of Bouma cycles. Most beds contained balls and pillow structures as well as load and scoured surfaces. Some mudstone beds show groove casts. Deformational structures are represented by slump and slide beds and sometimes convolute beddings.

Trace fossils are identified in sandstones, mudstones and muddy limestone beds. cf. burrows, borings, and trails.

Contacts: the lower is gradational with Tanjero formation and the upper part is conformable with Gercus formation.

Age: based on previous studies, the age of the Kolosh formation is Paleocene-lower Eocene.

Secondary features are (a) compaction of some mudstone beds and development of shale-like fissile, (b) development of large concretions, (c) bioturbation in parts, and (d) faulting.

5. Petrographic examination

Thirty-five thin-section slides are carefully studied to identify the mineralogical assemblages of the Kolosh sandstones. The modal analysis and average percentages are shown in **Table 1**. Petrographic identification of various mineralogical constituents is followed [21–28].

Olivine group: varieties of olivine minerals are identified, such as fayalite, forsterite, monticellite, and chondrodite, with an average percentage of 8.3% (Pl/1 A-E). They are mostly prismatic, angular to subangular, and are less common than pyroxene. Most olivine grains are fresh and few grains are uralitized and others have iron oxides and serpentine along the fractures.

Pyroxene group: Clino and orthopyroxenes are recognized based essentially on the extinction angle and petrographic properties. The average percentage of pyroxene minerals is 18.26%.

Grains composition	Average modal %	Grains composition	Average modal %
Pyroxene	18.3	Alkali Feldspar	1.0
Olivine	8.3	Quartz	2.5
Igneous rock fragments	11.3	Cristobalite	0.5
Altered fragments (Fe-Mg minerals)	17.4	Chert and chalcedony	2.5
Plagioclase-feldspar	5.2	Argillaceous fragments	3.5
Serpentinite MRF	4.4	Carbonate rock fragments	2.25
Opaque minerals (include chromite and glauconite)	4.0	Argillaceous matrix	16.5
Chlorite	1.6	Carbonate cement	1.25
Sum	70%		30%

Table 1.
The average percentages of the mineralogical constituent in the studied sandstones of the Kolosh formation in Badelyan locality.

Clinopyroxene is of hedenbergite, aegirine-augite, diopside, spodumene, pigeonite, and jadeite minerals. These grains are mostly fine and angular to subangular and of prismatic shape (Pl/1 F and H, Pl/2 A-D). Most of the clinopyroxene grains are fresh and others are altered to iron oxide or uralite in the center and a rim of pyroxene is still fresh (Pl/3 G).

Orthopyroxene: hypersthene and enstatite are recognized (Pl/1 G, Pl/2 D), mostly prismatic, angular to subangular, fresh with few altered grains. Orthopyroxenes are less common than clinopyroxene grains.

Feldspar group: plagioclase and alkali-feldspars are identified such as Ca-plagioclase as major type mainly of anorthite and minor bytownite and labradorite composition. The alkali-feldspar is anorthoclase and sanidine (Pl/4 A-D). Feldspar grains are mostly prismatic, fine, angular to subangular and have about 6.17% altogether. Some grains show sericite or carbonate replacement but keep the original shape.

Quartz attains average percentage of 2.05% of total grains, and has fine and subangular to subrounded shape (Pl/4 F).

Cristobalite attains average of 0.51%, less than quartz grains and is fine, angular to subangular grains (Pl/4 E).

Rock fragments: various types are recognized mostly igneous types with subordinate sedimentary and rare metamorphic fragments. The types of igneous fragments are ultrabasic, basic and few acidic, and intrusive and extrusive grains. While the types of sedimentary fragments are argillaceous, chert, chalcedony, and carbonate grains. The average percentage of various types of rock fragments is 15–20%.

Igneous rock fragments of intrusive and extrusive types, are composed of mainly ultrabasic, basic, and subordinate acidic fragments, attaining average of about 11.35%, these are:

- Intrusive igneous rock fragments varieties are ultrabasic igneous grins, including peridotite, heirzburgite, lherzolite, spilite, and pyroxenite (Pl/2 E-H, Pl/3 B-D). The basic intrusive fragments are composed of gabbroic grains (Pl/3 H). These grains are fine with few medium-grained, angular to subangular, and fresh with few altered grains. Some ultrabasic grains are serpentinized along the fractures or altered crystals in the fragment. Basaltic and andesitic grains of extrusive type are recognized also (Pl/3 E). These grains are angular to subangular and fine and mostly fresh and unaltered.

- Granitic fragments are acidic intrusive type, fresh, fine, and angular to subangular grains (Pl/3 F).

- Serpentinite grains attain average percentage of 4.42%, and are mostly fresh and are subangular to subrounded shape (Pl/3A).

- Chlorite grains attain average percentage of 1.56%, and are mostly of subrounded shape (Pl/4 G). Chlorite grains are fine-grained and have a dark green color.

- Altered grains are relatively of high percentages and attain an average of 17.4%. The grains of olivine, pyroxene, and ultrabasic and basic fragments show various degrees of alteration, mostly to iron oxides and/or uralitized pyroxene (Pl/4 A and C). The grains are altered to brown and dark brown in the center and the edges are still fresh. These are fine and subangular to subrounded grains.

- Sedimentary rock fragments are composed of limestone (includes fossils) and argillaceous fragments (Pl/3 H). The average percentage is 8.25% and is fine and subangular to rounded grains.

- Chert and chalcedony have an average percentage of 2.3%, are angular to subangular and composed of microcrystalline quartz (Pl/3 H). Chalcedony grains are rounded, composed of radiating silica, and sometimes filled with opal-CT. Radiolarian fossils are identified filled by chert and/or chalcedony.

- Opaque minerals include the majority of magnetite, chromite, chrome-spinel, and hematite. The average percentage is 4.0%, and are fine to very fine, angular to subangular in shape.

- Glauconite grains are scattered in the sandstone units of mostly rounded shape and green color.

- Cementing materials: Argillaceous matrix is the main type of cement with few carbonates. The percentages of the muddy matrix are more than 16.53% suggesting greywacke sandstone type [13], while carbonate is 1.2%.

The classification of sandstone cf. [13] and [14] provenance and tectonic setting diagram cannot applicate due to the high content of olivine, pyroxene, ultrabasic and basic, and altered fragments, which reach more than 50%.

6. Facies analysis

Classification of facies types in the Kolosh formation is followed [15–20]. Sedimentological field studies combined with microscopic facies analysis of Paleocene siliciclastic sequences in the Badelyan area have resulted in the recognition of 11 sedimentary facies (**Table 2**, plate/5). Facies types are grouped into four main categories, based on their lithological characteristics and formation processes, comprised mainly of mudstones, sandstones, and minor carbonate and silicified facies (**Table 2**). The textural characteristics, grain and facies association and depositional settings of the studied rocks are summarized in **Table 2**.

Firstly, facies description and environmental interpretation of the facies are discussed. Further details on sedimentary environments and regional depositional settings are provided in the following;

Eleven lithofacies types are composed of interbedded thick mudstone, muddy sandstones, and pebbly sandstones arranged basically in fining upward Bouma turbidity cycles [29]. Bouma cycles are capped with pelagic limestone or silicified mudstones. The field observations of facies types and sedimentological features are compatible with deep water slope apron to basin plain depositional system. Specific facies types of turbidite origin are discussed and interpreted in **Table 2**.

6.1 Facies association

Lithofacies types of the Kolosh formation are grouped into four main facies associations, each one is referred to a certain depositional system, these are:

Facies	Thick (m)	lithology	Characteristics	Process
F1	0.25–0.5	Laminated/fissile shale	Planner to undulated, laminated dark gray to black silty shale, 0.5–1.5 cm thick laminates	Hemipelagic
F2	0.25–0.5	Graded bedded muddy sandstone	Thin beds fining upwards, arranged in turbidity cycles, fine to very fine sandstones	Middle to outer submarine fan turbidite
F3	0.25–1	Calcareous mudstone	Dark gray, graded lower base and sharp upper base	Hemipelagic/pelagic mud-low energy turbidite
F4	1–10	Deformed slump and convolute sandstone/ mudstone	Slump, slide, and convolute beds of sandstone and mudstone sometimes a mixture of dark gray to green color, scour and fill structure	Submarine fan channel levee High-energy turbidity currents
F5	0.15–0.5	Undulated and wedge sandstone/ mudstone	Wedge beds of dark green sandstone and gray mudstone, undulated beds	Edge of submarine fans, low energy turbidity currents
F6	1	Muddy pebbly sandstone	Dark green, coarse-grained turbidite, sharp, scour, and grooved surface	High concentration turbidite/ debrite
F7	0.5–1	Silicified mudstone/chert	Discontinuous or lenticular in part, pale gray, silicification	Pelagic mud to very low energy turbidity currents
F8	1–2	Chaotic mud-rich with hard sediments of muddy sandstone and calcareous mudstone	Disturbed, chaotic-hard mudstone and sandstone, deformed, load, scour surfaces, green to gray color	High energy turbidity currents, slope submarine fans-channels
F9	1–10	Submarine channel sandstone	Fine to very fine-grained, scoured, load and flute casted surfaces, dark green color	Slope submarine fan turbidite, low concentration turbidite
F10	1–1.5	Muddy pebbly bioclastic sandstone	Sharp base, scoured	Bioclastic rich turbidite
F11	0.1–0.2	Fossiliferous muddy wackestone	Continuous to discontinuous wispy beds with gradational contact	Pelagic/hemipelagic carbonate

Table 2.
Lithofacies types identified in the Kolosh formation of Badelyan locality.

6.1.1 FA1, slope/submarine-channel fan

The lower part of the Kolosh formation in the studied area represents submarine channels developed in slope margins. Submarine erosion and deposition of the sediment is performed by gravity flows and/or turbidity currents as a result of migrated active channel floor [20, 30–32]. Debrites and developed turbidites comprise two stages of the sediment-gravity-flows. The debris flows of muddy pebbly sandstone are represented by suites of facies F4, F5, F6, F8, F9, and F10. The coarse grains and gravel are supported by a cohesive matrix of interstitial fluid and mud [20]. Sand and mud turbidites are deposited from suspended loads and graded upwards by fluid turbulence in the inner channel and on levee. Channel sediments commonly deposited the following facies, represent a spectrum of submarine mass-movement processes [20, 33–39].

1. Thick-bedded, amalgamated sandstones, and/or pebbly sandstones are deposited by collapse of high-density turbidity currents through traction reworking of bed load sediment (cf. F4, F6, F8).

2. Interbedded thin fine sandstones/mudstones are deposited from low-density turbidity currents (cf. F2, F5, F7).

3. Thin bedded and stratified mudstones/shales, deposited from dilute, low-density turbidity currents and subsequent suspension are deposits of mud between turbidity currents (cf. F1, F2).

4. Nongraded sandstone and/or chaotic mud-rich facies cf. F8, with harder sediment of F1/F2 facies within a mud-rich matrix, deposited from debris flows.

5. Deformed, contorted (overturned and/or offset stratification) heterolithic horizons deposited from slumps and/or slides (cf. F4, F5, F7) [39].

6.1.2 FA2, inner fan/channel levee

The sandstone/mudstone beds are arranged in turbidity Bouma cycles with or without basal division (Ta). The thickness of these beds ranges from 0.5 m for sandstones and 1.5 m for mudstones. Balls and pillows, disturbed beds, convolute beds, slump and slide beds, load, and scour and grooved surfaces are the main characteristic sedimentary structures.

Slump and/or slide of interbedded sandstone/mudstone horizons on both sides of the channel valley are filled and interfinger with channel and interchannel deposits that are interpreted here as levee/inner fan deposits cf. F2, F3, F4, F5, F7. Deformation of the slump is represented by recumbent folded horizons bounded by undeformed beds. The sandstone's and mudstone's balls and pillows are associated with slumping and/or sliding strata, and some are very large with a diameter of about 1 m.

6.1.3 FA3, outer fan

The outer fan deposits consist of thick mudstone/shale beds interbedded with very thin sandstone horizons graded upwards to reveal Bouma cycles without the basal divisions (Ta, Tb). The outer fan succession comprises a developed thick sequence of alternating non-channelized sandstone bodies and associated thin/thick bedded argillaceous mud sediments cf. F1, F2, F3, F5, F7, F11.

The middle part of the Kolosh succession cf. 45–90 m above the lower base permits detailed inspection of numerous thickening upward cycles. Each cycle consists of lower thin-bedded and fine-sandstone facies, transitionally overlain by thicker bedded/massive mudstones. These criteria and terminology are already discussed by Refs. [30, 31, 40, 41], which are interpreted as lobe-fringe and lobe-fan deposits, respectively. These two facies' cycles are either symmetric or asymmetric, thus reflecting sudden or gradual shifting of the related feeder system as suggested by Refs. [20, 41].

Typical sedimentological features of thin-bedded lobe-fringe deposits are included cyclic vertical variations depicted by changes in the sand/shale ratio and in the sandstone bed thickness.

The lobe-fringe turbidite beds displayed base-missing of Bouma sequences, that is, Ta, Tb, and/or Tc. Typically, by far, the Tc-e subdivisions are predominant. In part, Tb-e cycles are displayed unusual thick beds, which are randomly scattered in the succession. The sandy beds' division, which includes current-ripples is bounded by upper surface gradational with the overlying turbiditic mudstone.

The post-depositional and/or syndepositional plastic deformation and liquefaction are the causes of abundant convolute, slump, balls, pillows, and the small-scale features, such as pseudo-nodules observed, in the sandy and muddy portions of many beds. Occasionally thin veneers of bioturbated hemipelagic mudstone and/or lime mudstone are found at the top of some turbidity cycles.

6.1.4 FA4, pelagic/hemipelagic basin plane

Hemipelagic/pelagic basin plain sediments of the Kolosh formation are very monotonous sequences of alternating thin sandstone, siltstone, mudstone, and hemipelagic silicified mudstones. These beds are associated with numerous intercalations of thin turbidity

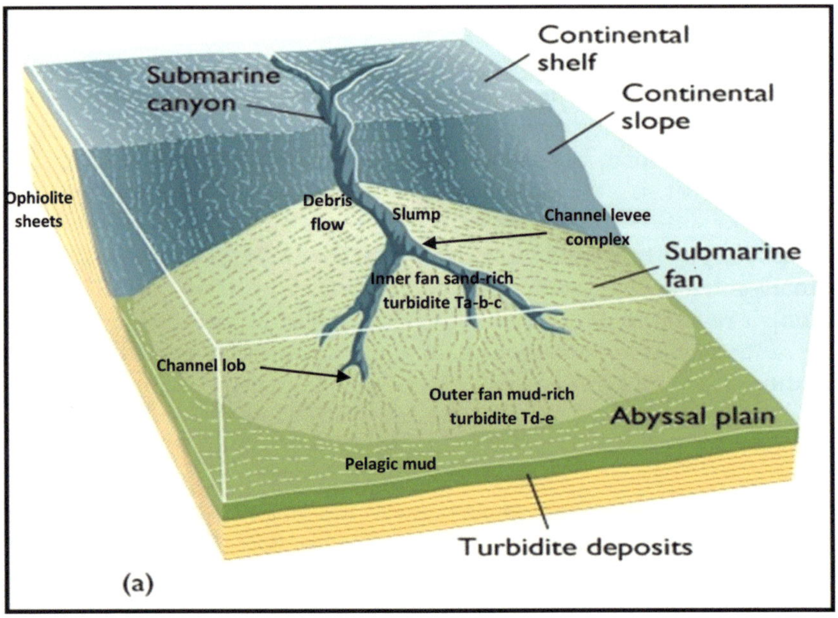

Figure 3.
Depositional model of the Kolosh Formation discusses the slope and basin plan with anatomy of submarine fan and lithofacies types. Ta-b-c-d-e are subdivisions of Bouma turbidity cycle.

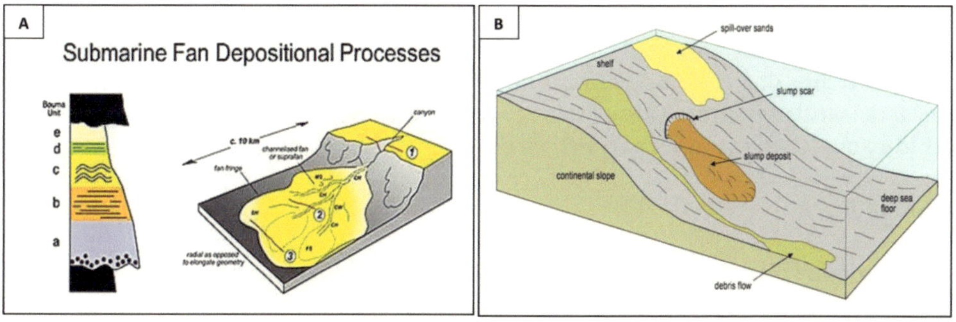

Figure 4.
Schematic anatomy model shows subdivisions of the submarine fan and Bouma turbidity cycle; A. Bouma cycles subdivisions, Ta to Te e.g. Ta= pebbly to coarse-grained sandstone, Tb= laminated sandstone, Tc= cross-bedded sandstone, Td=very fine sandstone to siltstone, Te=mudstone/shale and position in the fan. And subdivisions of submarine fan; 1) canyon, 2) Mobile channels with wings, 3) fan fringe and abyssal plan B. Facies types in the slope and submarine fan; slump deposits, debris flows and deep sea floor cf. abyssal plain (adopted from Mutti [16]).

lime mud beds and related slump units, which belong to quite distinct depositional settings [16, 30, 31, 42]. The carbonate mud/marl beds are derived from tectonically unstable flanking platforms (cf. around island arc) and include a variety of fine-grained distal turbidites facies rich in planktonic foraminifera and radiolarian fossils.

The basin plain sediments of the Kolosh formation are characterized by the following features:

The surfaces of the beds are even parallel regardless of the extent of exposures and turbidite beds are missing the base of Bouma cycles with predominant Tc-e divisions. The sandy and/or silty division of the turbidite beds are of thin parallel laminae (Td-division), which grades transitionally into turbiditic mudstone (Te-division) (**Figures 3** and **4**). Typically, the small-scale cross/wavy laminae of the current-ripple division (Tc) display low to very low angles and occur in well-developed climbing patterns. The basal portion (Tc) is fine to very fine sandstones. The turbidite beds are regularly alternating with thinner and bioturbated hemipelagic beds. Moreover, these beds contain distinctive assemblages of indicative benthic foraminifera of water depths in the range of 600–2000 m, as well as planktonic types [16].

7. Tectonosedimentary evolution

The depositional evolution of the Kolosh formation was mainly controlled by continuous subduction of continental and oceanic crust in the Zagros Tethyian basin and was indirectly related to the sea level changes. The Kolosh successions can be classified into four stratigraphic sequences, representing the certain depositional system. Successive facies associations are arranged from bottom to top cf. slope/submarine-channel fan FA1, inner fan/channel levee FA2, outer fan FA3, and pelagic/hemipelagic basin plane FA4 refer to progressive deepening from slope apron to the basin plain in the top of the formation (**Figures 3** and **4**).

7.1 Stage 1

The first tectonosedimentary stage comprises slope/submarine channels that flows down the slope and comprised of F4, F5, F6, F8, F9, and F10. These are evident from the bed geometry, sedimentary structures, and facies types. FA1 is characterized by muddy, pebbly coarse-grained turbidite sandstone F6, chaotic mud-rich facies, with harder horizons of F1/F2 facies in a mud-rich matrix F8, and muddy, pebbly, and bioclastic-sandstone F10. These facies are typically originated from relatively dense turbidity currents, in slope margins via the submarine channel.

7.2 Stage 2

Continuous thrusting developed subduction and consequently transgression and deepened into the forearc basin. These formed distributary channels in the inner margins of submarine fans, which are evident by deformational structures, such as slump, convolute, and channel levee beddings of F4 and F8, composed of thick/thin interbedded mudstones and sandstones.

7.3 Stage 3

Continuous subduction and transgression enhanced deepening and shifted the sediments inward the basin. Sedimentation was developed from the slope apron margin cf. FA1 and FA2 to the outer submarine fans in the basin plain. FA3 is composed of thin/thick-bedded argillaceous muddy sediments intercalated with thin beds of sandstone and deep marine carbonate mud rich in planktonic foraminifera and radiolarian fossils cf. F2, F3, F5, F7, and F11, thus refer to deepening in the Kolosh basin.

7.4 Stage 4

Continuous subduction and transgression shifted the sediments to the pelagic/hemipelagic apron, representing the last stage of tectonosedimentary evolution of the Kolosh forearc basin. These are evident from suite of facies types of alternative thin beds of sandstone, siltstone, mudstone, and hemipelagic carbonate mud/marlstone cf. F1, F2, F3, F7. F11. Moreover, it is evident from bed geometry, sedimentary structures, and microfossils assemblages, such as radiolaria, planktonic foraminifera, and hemipelagic benthic foraminifera. This margin is characterized by silicified mudstone and discontinuous silica horizons.

8. Discussion

8.1 Mineralogy of the source rocks

At Badelyan locality, Kolosh and Tanjero formations are exposed at the suture boundary of the Iraqi thrust zone near Jabal Hassan Beg mountain. These are comprised of the accretionary prism lying directly beneath the emplacement sheets of thrust zone. Based on field observations, characteristic sedimentary structures and sedimentological features all together refer to deep marine sedimentation originated by turbidity currents.

Petrographic examination of the Kolosh sandstones reveals high percentages of fresh olivine, pyroxene, intrusive ultrabasic, and basic and extrusive basic igneous rock fragments, which are controlled by the provenance and tectonic evolution of the Zagros thrust belt and surrounding areas. The mineral assemblages reveal various potential tectonically-controlled source rock complexes (cf. mantle, oceanic crust, continental crust, and volcanic arcs). The presence of olivine, diopside, and augite accompanied with ultrabasic rock fragments (cf. dunite, peridotite tectonite, heirzburgite, lherzolite, and pyroxenite) suggest ophiolitic mantle origin [21, 23–26, 28, 43].

The peridotites of the ophiolite belt are ranging in composition from lherzolite to dunite through heirzburgite. These rock types and mineral associations are similar to those of forearc peridotites, which are represented by fertile alpine mantel lherzolite to dunite and are depleted in tectonite heirzburgite [44]. It is suggested that these ultramafic bodies are huge fragments of supra subduction zone of residual mantle peridotites [45]. Serpentinization is observed in along cleavages and basal fractures of the Fe-Mg minerals [25], as complementary processes during the evolution of the ultramafic part of the ophiolite sequence adjacent to the foreland basin. Serpentinization of peridotite is thought to have taken place during the subduction stages before the collision of the Arabian plate with the Asian plate [21, 45].

Diopside-diallage is usually derived from coarse-grained gabbro and basalt and is usually accompanied by forstertic olivine [44]. Omphacite is similar to diopside, augite, and jadeite, which are found exclusively in eclogite cf. deep mantle of high T and P conditions. Hedenbergite is usually derived from peridotite and fayalite ferrogabbro [25].

Augite is essential mineral in peridotite, gabbro, basalt, and olivine gabbro. Pigeonite may be derived from basalt, diabase, and dolerite, whereas, aegirine-augite compose an essential member in the trachyte and basalt. Spodumene is a rare mineral, derived from Li-granite [25]. Enstatite occurs in all types of basic igneous rocks. Mg-rich orthopyroxene occurs in ultrabasic igneous rocks (cf. pyroxenite, heirzburgite, lherzolite, serpentinite, and picrate with Mg-olivine) [25]. Hypersthene is found in gabbro, norite, and andesite [21, 26, 27].

Mg-rich olivine (e.g., forsterite and monticellite) is an essential mineral in most ultrabasic igneous rocks (cf. dunites, peridotites, and picrites). Olivine forsterite and chrysolite are derived from peridotite, olivine gabbro, and basalt. More Fe-rich olivine (e.g., fayalite) is derived from alkali basalt, ferrogabbro, and trachyte [25]. Monticellite is derived from basalt, and chondrodite is derived from high-grade metamorphic rocks [21–23, 26, 27].

The altered grains show various degree of alteration and are found in relatively high percentages. It is composed mainly of uralite and iron oxides. Most of the altered grains are derived from basic and ultrabasic rock fragments as well as olivine and pyroxene.

Ca-plagioclase members An95-An55 cf. anorthite, bytownite, and labradorite, are characteristic minerals of ultrabasic and basic igneous rocks cf. peridotite, heirzburgite, lherzolite, gabbro, and basalt. Anorthite is derived from ultrabasic intrusive igneous rocks cf. peridotite, pyroxenite, heirzburgite, and lherzolite and from gabbro. Bytownite and labradorite are derived mainly from basalt and gabbro [23, 26, 43, 44]. The alkali-feldspar varieties are derived from acidic intrusive and extrusive igneous rocks. Orthoclase is derived from granites, while anorthoclase and sanidine with cristobalite are derived from acidic and intermediate extrusive igneous rocks, such as rhyolite, trachyite, and andesite [24–26, 43, 44, 46].

Varieties of rock fragments are important indicators of source rocks and provenance. The most important abundant types in the Kolosh sandstones were derived from intrusive ultrabasic and intrusive and extrusive basic igneous rocks (cf. pyroxenite, lherzolite, heirzburgite, peridotite, gabbro, and basalt). These rock types are derived from mantle and oceanic crust as an ophiolite sequence [21, 25, 26, 44]. The basalt and andesite fragments are derived from volcanic arcs [26, 44].

The sedimentary rock fragments were derived from various sources. Intrabasinal mudstone grains are most probably derived from mudstone beds by the effect of turbidity currents. The limestone fragments are extrabasinal and derived from Cretaceous carbonate formations. The chert and chalcedony fragments are suggested to derive from acidic volcanic rocks of arc and/or deep marine radiolarian beds of the ophiolite sequence [17, 18, 20, 44].

8.2 Potential ophiolites provenance

Anomalous concentrations of pyroxene, olivine, and ultrabasic/basic igneous grains in the Kolosh sandstones are suggested to derive from two sources: thrusted ophiolite sequences and lava flow basalt from island arc in the foreland basin. The emplacement of the Neotethyan oceanic crust took place during late Cretaceous (118–97 Ma), while the emplacement of Hasan Beg ophiolite complex was (106–92 Ma) ago [47]. Along the southern Neotethys suture, the outer Zagros orogenic belt (OZOB) crops out, including Rayat, Piranshahr, Kermanshah, Neyriz, and Haji-Abad Ophiolites resulted, of the late Cretaceous collision between the Sanandaj-Sirjan and the Arabian shield [47, 48].

Sedimentation of the qulqula group cf. deep-marine radiolarian chert, and carbonates of Arabian passive-margin is dated as Valanginian-Turonian (140–89 Ma), which constrains initial orogenesis and early subsidence of the Zagros foreland basin to about 90 Ma [49–51]. The provenance data of the clastics in the Tanjero formation (Maastrichtian), Kolosh formation, and Suwais red beds (Paleocene-Eocene) reveal partial derivation from basic and ultrabasic sources related to ophiolite emplacement [49–51]. In the studied area, several pre-tertiary tectonically-derived igneous complexes may have served as source rocks for the sandstones of the Kolosh formation.

Hasan Beg mountain successions is situated beside the Badelyan area and very near to Sidekhan province. Hasan Beg igneous complex comprises a late Cretaceous remnant of the ophiolite-arc system that developed within the Neotethys ocean. This igneous complex was subsequently accreted to the Arabian plate during the late Cretaceous to Paleocene [50]. It is predominantly consisting of calc-alkaline basaltic andesite to andesite cutting across by micro gabbro and diorite dikes indicating Albian-Cenomanian age (106–92 Ma) [47, 50]. Hasan Beg mountain rock units start at the bottom by pillow lavas with sheared and highly weathered intense deformed chlorite slate, which forms the contact between the lower pyroclastic metavolcanic rocks and the upper part of metasediments cf. slate, shale, and sandstone. The overlying metasediments are composed of highly fossiliferous black shale interbedded with 20 m thick sandstone lens. These rocks are overlain by radiolarian chert with a total thickness of about 1000–2000 m, and this variable is due to thrusting [52].

Qalander-sidekhan mountain sequence is located about 20 km from Badelyan locality and composed from bottom to top of Tanjero and Aqra formations (late Cretaceous) separated by tectonic contact from the overlain govanda limestone (early-middle Miocene age), succeeded by red bed sandstones (middle-late Miocene), which overlies by a tectonic slice of naopurdan metavolcanic rocks [52]. In the Qalander locality the naopurdan rocks are divided into three parts: 1) The lower metavolcanic unit is composed mainly of basalt flows, 2) The middle unit is variable in thicknesses averaging about 50 m and composed of pillow lavas, brecciated lavas, and inter-pillow ash. Many dolerite dykes are cutting across the Qalander successions, in which pillow and flow lavas are about 450 m thick. The upper unit contains sequences of sandstones grades upward to conglomerates, containing pebbles of basalt, serpentinite, and marble, which have been derived from erosion and rapid deposition of the igneous rocks, probably from naopurdan and walash metavolcanic successions [45, 53].

The Sidekhan mountain, located NW of Qalander mountain has the same successions as the Qalander mountain with govanda formation and naopurdan successions, with tectonic breccia.

Rayat ophiolite mantle sequence lies in the NE corner of Iraq (30 Km from Badelyan area) and consists of serpentinized peridotites, serpentinite, tectonite heirzburgite, lherzolite, and dunite with Cr-spinel [27]. They are mantle residues with distinct geochemical signatures of affinities of both the mid-ocean ridge and supra subduction zone. The crustal succession includes gabbro, dibasic dikes, rare pillow basalts, and radiolarite overlain by late Cretaceous pelagic limestone. Rayat ophiolite is extended to Piranshahr area in NW Iran and to the Cilo ophiolite in SE Turkey [44, 46]. The Rayat ophiolitic mélange is mainly composed of peridotite and sheared serpentinite with heirzburgite [45]. The chromitite in the Rayat peridotite is similar to the mantle chromitite and Moho transition zone chromitite (upper mantle zone) of the Tethyan ophiolites [45].

Galalah-choman ultramafic/mafic rocks are situated about 25–28 km from Badelyan area. Galalah area consists of 50 m brecciated and pillow lavas alternating with ultramafic/mafic volcanoclastic rocks, and above with 40 m massive serpentinite and mélange marble of heterogeneous composition (mixture of calcite, chlorite, and serpentinite). This succession is overlain by a thin layer of radiolarian chert and red beds, which are tectonically separated by crush zone [54]. Choman successions are variable in thicknesses of about 200 m of pillow lava, interbedded with volcanic ash, tuff, and breccia [45]. Generally, galalah and choman rocks consist of basaltic flow lava interbedded with ultramafic/mafic rocks, overlying serpentinite rocks. They can be suggested as other sources of pyroxene, olivine and ultrabasic, and basic intrusive and extrusive basaltic igneous grains in the sandstone units.

The kata rash igneous complex comprises Cretaceous remnant arc and is broadly similar in age to those of late Cretaceous peri-Arabian ophiolite belt in other countries as well as other late Cretaceous Zagros supra subduction zone assemblages [53, 55]. This refers to the great lateral extent of late Cretaceous arc systems in the consumption of the Neotethys ocean.

8.3 Sedimentary mechanism of olivine/pyroxene concentration

Ophiolite successions are suggested as the main source of pyroxene and olivine grains in the Kolosh sandstones. It consists of rocks of various pyroxene/olivine-rich assemblages, including peridotite tectonite, lherzolite, serpentinized spinel peridotite, pyroxenite, serpentinite chromitite, heirzburgite, gabbro, and basalt lavas.

Later, slumps of these pyroxene/olivine-rich sands by turbidity currents are transported deeper in the basin to deposit finally as a turbidity sequence. Tectonic activity and seismicity in the foreland margin usually create turbidity currents and slumping of accumulated heavy sarends in the shallow zone to deeper margins [7, 16, 56].

Zagros Cretaceous ophiolite complexes are exposed in the early tertiary and are amenable to subaerial and shallow submarine erosion along the shorelines. Intense subaerial erosion enhanced the rapid mechanical disintegration of the ophiolites evidenced by the fresh olivine and pyroxene grains in the studied sandstones.

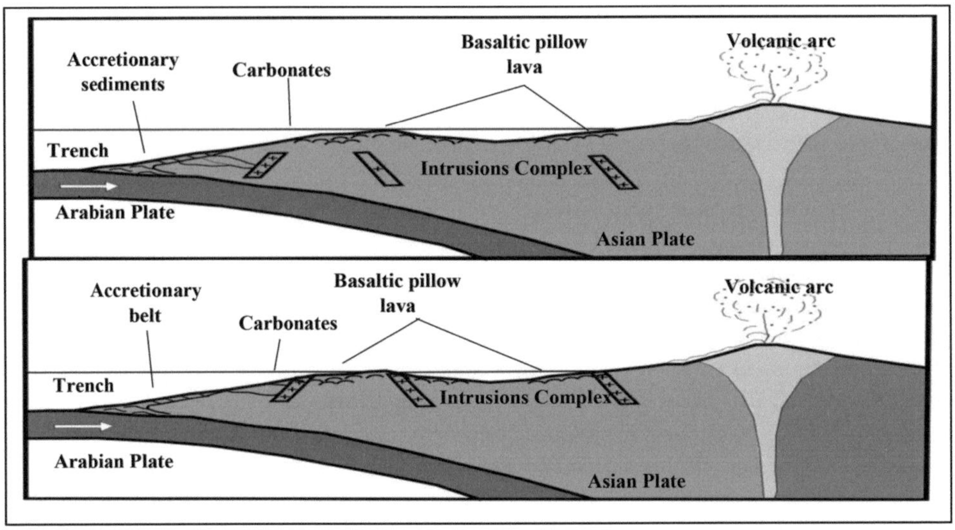

Figure 5.
A model discusses the Late Cretaceous geodynamic evolution of subduction Arabian and Asian plates to form Zagros foreland basin, in Late Cretaceous-Paleocene age A) Various intrusive magmatic bodies with arc-related basaltic flows, B) continuous subduction and subsides of the Kolosh basin (trench) and subsequent erosion of ophiolitic; ultrabasic, gabbro, to deposit the Kolosh Formation in the continental slope and/or deep marine trench (modified after Calvo [57]).

Submarine wave action in shallow margins induced concentration and accumulation of the heavier fraction of the detrital rich in pyroxene and olivine grains according to the model discussed by [57] (**Figure 5**).

The Kolosh formation is suggested to deposit in the deeper Zagros trench subduction margin. Intense tectonic activity and/or earthquake seismicity are responsible for successive episodic turbidity currents that deposited the Kolosh successions.

8.4 Depositional environment and evolution

Grain sizes, sedimentary structures, and facies types indicate tectonosedimentary evolution of deep marine environment (**Figure 3**). The facies types and associations refer to slope submarine channels advanced to submarine fans in deeper margins. The facies types accompanied by sedimentary structures refer to a debris flow in the inner fan advanced to outer fan and later to basin plane pelagic/hemipelagic margins (**Figures 4** and **5**). Submarine fans include lobe-fans and lobe-fringe produced by advanced turbidity currents toward the basin plane. Thin-bedded successions of the lobe-fringe are products of waning and relatively dilute turbidity currents.

The field observations show that the vertical and lateral extent of thin-bedded lobe-fringe sediments are the distal equivalents in both down-current and cross-current directions of the fine-grained and thin-bedded sandstone horizons that comprise the depositional lobes of the end of the outer fan margin.

Thus, the heavily concentrated turbidity currents, which reached the outer fan are dropped to the coarsest suspended load firstly in the inner fan margins. Through successive events, sediments are formed lobs of the coarse-grained and thick-bedded sandstone bodies, which are most probably deposited in and around the channel levee in inner fans. Farther downslope, the same current became more dilute suspensions carrying the finer-grained material to accumulate as peripheral fringes around the lobes. These turbidites are characterized by the down current areas from those reached by the lobes, which are discussed as fan lobe fringe.

Thin-bedded horizons of the outer fan and fan-fringe margins are undergoing alternative phases of high and low rates of sedimentation. The high rate is discussed by the active progradation of relatively sand-rich deposits, which is evident by the abundance of load and scour structures, particularly pseudo-nodules. These structures require soft pelitic sediments, which are rapidly overlain by denser sand beds. The latter horizons are represented by thin veneers of highly bioturbated hemipelagic mudstone/marl, which are found particularly in the thinnest and mostly shale facies of the lobe-fringe and fan-fringe deposits.

All features of the thin-bedded horizons in the Kolosh basin plain are result from the transport and sedimentation of sand, silt, and mud across the flat basin floor. Sedimentation of thin-bedded deposits is carried out by vanning, dilute/low-density turbidity currents that gradually lose the suspended load with distance. The rate of hemipelagic deposition between successive turbidite layers is apparently controlled by the distance from the source of the turbidity currents or, in other words, by the rate of turbidite sedimentation. Thus, hemipelagic mud interbeds appear to be increasingly thicker in a down-current direction compared to the turbidite dispersal system.

8.5 Tectonosedimentary evolution

Depositional evolution in the Kolosh foreland basin is extremely controlled by tectonic activity. The collision between Arabian and Eurasian Plates is started in late Cretaceous age accompanied by volcanic arc eruptions (**Figures 5** and **6**).

Continuous collision formed subsidence and deepening to create a trench basin and consequently transgression. The lower part of the Kolosh formation is deposited in the marine slope margin, which is evident from the sedimentological features supported by sedimentary structures and facies types. The identified sedimentary structures and related facies types in the middle part of the Kolosh formation suggest progressive development of the environment to inner submarine fan and channel

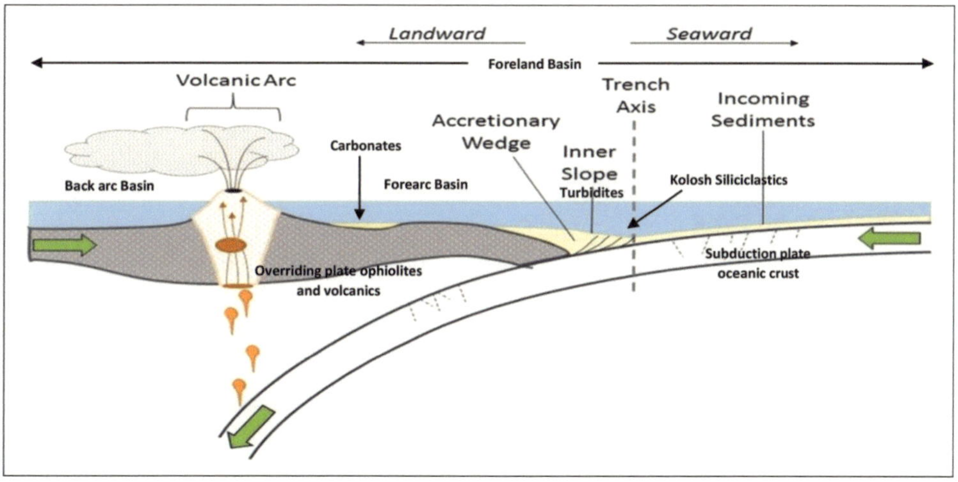

Figure 6.
Schematic diagram of foreland basin with forearc and back arc divisions and subduction zone. The diagram shows the site of deposition of the Kolosh Fm. In the accretionary wedge.

levee margins. Continuously thrusted evolution progresses the deposition to outer fan margins. The later stage of sedimentary evolution represents progress into basin plain margins and continuous collision of continental and oceanic crust creating a trench basin, where Kolosh formation was deposited.

9. Conclusions

The anomalous concentrations of fresh olivine, pyroxene, and ultrabasic/basic rock fragments grains in the Kolosh sandstones were derived from wide varieties of ultrabasic/basic intrusive and extrusive igneous rocks of pre-tertiary ophiolites from Zagros thrust zone. Olivine/pyroxene-rich sandstones of the Kolosh formation are important to study the regional correlation of the ophiolite's composition and the differences along the Neotethyan ophiolite belts of the Neotethys ocean.

Olivine and pyroxene grains were derived from varieties of rocks, such as tectonite peridotite, pyroxenite, lherzolite, heirzburgite, basalt, and dolerites, basically derived from Rayat and Hasan Beg ophiolites, the nearest to Badelyan study area. The radiolarian chert and carbonate formations with ophiolite rocks in Rayat, Choman, Galalah, and Hasan beg are suggested as sources of chert and carbonate rock fragments in the Kolosh sandstones.

The fresh olivine and pyroxenes grains are the products of subaerial and submarine erosion of the thrusted mantle and oceanic crust sheets. Intense wave action accelerated erosion on the beach and concentrate the heavy olivine and pyroxene grains, which were slumped to a deeper trench basin by the tectonically-induced turbidity currents.

Facies analysis and related sedimentary structures suggest progressive development in the Kolosh basin starting from slope marine margin in the lower part up to the basin plain in the upper part of the formation passing through the submarine fan environment.

The Neotethys foreland basin is tectonically controlled by thrusting and the Kolosh sediments are deposited in progressively deepening upward setting. According to the field observations, facies and associations and stratigraphic successions suggest four stages of tectonosedimentry development, these are slope channel, inner fan, outer fan, and hemipelagic/pelagic apron. These stages are controlled directly by tectonism and seismicity created by continuous thrusting.

Author details

Al-Mashaikie Sa'ad ZA Kader
Department of Geology, College of Sciences, University of Baghdad, Al-Jadryiah, Baghdad, Iraq

*Address all correspondence to: magnesite2006@gmail.com

References

[1] Sissakian VK, Fouad SF, Danyal SY. Geological Map of Arbeel and Mahabad sheet NJ-38-14 and NJ-38-15. Baghdad, Iraq: GEOSURV; 2014

[2] Al-Mashaikie S, Mohammed fI J. New discovery of olivine/pyroxene-rich sandstones in the Kolosh formation, Dyana area, NE Iraq: An approach to provenance and tectonics of the sedimentary basin. Iraqi Bulletin of Geology and Mining. 2018;**14**(2):81-96

[3] Ditmar, V., Iraqi–Soviet Team, 1971. Geological conditions and hydrocarbon prospects of the Republic of Iraq (northern and central parts). Manuscript Report. (Iraqi National Oil Company) Baghdad: INOC Library.

[4] Buday T. In: Kassab IIM, Jassim SZ, (in collaboration with staff members and experts of the State Organization for Minerals, Baghdad), editors. The Regional Geology of Iraq, Volume 1, Stratigraphy and Paleontology. Directorate General of Geological Survey and Minerals; 1980. p. 445

[5] Jassim SZ, Goff JC. Geology of Iraq. Dolin Prague and Moravian Museum: Czech Republic, Brno; 2006. p. 341

[6] Perrone V, Perrotta S, Marsaglia K, Staso A, Tiberi V. The oligocene ophiolite-derived breccias and sandstones of the Val Marecchia Nappe: Lights for paleogeography and evolution of Northern Apennines (Italy). Palaeogeography, Palaeoclimatology, Palaeoecology. 2014;**394**(2014):128-143

[7] Renyck, H., Witmer, S., 2008. Olivine Beach on Big Island of Hawaii. Available from: http://epod.usra.edu/blog/2008/01/unknown-1302008.html

[8] Dunnington HV. Generation, migration, accumulation and dissipation of oil Northern Iraq. In: Weeks GL, editor. Habitat of Oil. Tulsa: American Association of Petroleum Geologists; 1958. pp. 1194-1251

[9] Jassim SZ, Al-Shaibani SK, Ajina TM. Possible middle eocene block movement in Derbendikhan area, northeast Iraq. Journal of Geological Society of Iraq. 1975;**S**(1):139-145

[10] Kassab IM. Planktonic foraminiferal range in the type Kolosh Formation (Middle-Upper Paleocene) of Northeast Iraq. Journal of the Geological Society of Iraq. 1976;**4**(special issue)

[11] Bellen RCV, Dunnington HV, Wetzel W, Morton DM. Lexique Stratigraphique International, Asie Fasc. 10a, Iraq. Paris: Centre National de la Recherche Scientifique; 1959. p. 333

[12] Whitney DL, Evans BW. Abbreviations for names of rock-forming minerals. American Mineralogist. 2010;**95**:185-187

[13] Pettijohn FJ. Sedimentary Rocks. 3rd ed. New York: Harper and Row; 1975. p. 628

[14] Dickinson WR. Interpreting provenance relations for detrital modes of sandstone. In: Zuffa GG, editor. Provenance of Arenites, NATO ASI Series C. Vol. 148. Dordrecht: Reidel; 1985. pp. 333-361

[15] Selley RC. An Introduction to Sedimentology. London, New York, San Francisco: Academic Press; 1976. p. 408

[16] Mutti E. Turbidite Sandstone. Italy: Istituto di Geologia, Universit'a di Parma, Agip, S.P.A; 1992. p. 275

[17] Walker J, James A. Facies Models in Response to Sea Level Changes. Canada Geosciences: Geotext-1. Canada: Geological Association of Canada; 1998. p. 454

[18] Einsele J. Sedimentary Basins, Evolution, Facies and Sediment Budget. Berlin: Springer-Verlag; 2000. p. 628

[19] Stow AVD. Sedimentary Rocks in the Field a Color Guide. Germany: Academic Press, Elsevier; 2012. p. 320

[20] Pickering KT, Scott RN. Deep Marine Systems, Processes, Deposits, Environments, Tectonics and Sedimentation. United Kingdom: AGU and Wiley and Sons; 2016. p. 657

[21] Kerr FP. Optical Mineralogy. 6th ed. New York: McGraw-Hill Book Co.; 1975. p. 442

[22] Scholle PA. A Color Illustrated Guide to Constituents, Textures, Cements, and Porosities of Sandstones and Associated Rocks. Tulsa, OK: The American Association of Petroleum Geologists with the support of The American Association of Petroleum Geologists Foundation; 1979

[23] MacKenzie WS, Guilford C. Atlas of Rock-Forming Minerals in Thin Section. England: Pearson Prentice Hall, Pearson Education Limited; 1982. p. 98

[24] Adams AE, Mackenzie WS, Guilford C. Atlas of Sedimentary Rocks Under the Microscope. England: Longman Group Limited, Longman House; 1984. p. 104

[25] Gribble CD, Hall AJ. Optical Mineralogy Principles and Practice. Boca Raton, FL, London, New York: Taylor and Francis Group; 1992. p. 302

[26] Adams AE, Mackenzie WS. A Colour Atlas of Carbonate Sediments and Rocks Under the Microscope. Barcelona, Spain: Mason Publishing; 2001. p. 180

[27] Barker A. A Key for Identification of Rock-Forming Minerals in Thin Section. London, UK: Taylor and Francis Group; 2014. p. 181

[28] Ghose NC, Chatterjee N, Fareeduddin A. A Petrographic Atlas of Ophiolite, An Example from the Eastern India – Asia Collision Zone. New Delhi, Heidelberg, New York, Dordrecht, London: Springer; 2014. p. 234

[29] Bouma AH, Brouwer A, editors. Turbidities. Development in Sedimentology, 3. Amsterdam: Elsevier; 1964. p. 264

[30] Shanmugam G, Moiola RJ. Submarine fans: Characteristics, models, classification, and reservoir potential. Earth-Science Reviews. 1988;**24**(6):383-428

[31] Shanmugam G, Moiola RJ, McPherson RG, O'Connell S. Comparison of turbidite facies associations in modern passive-margin Mississippi fan with ancient active-margin fans. Sedimentary Geology. 1988;**58**(1):63-77

[32] Sylvester Z, Pirmez C, Alessandro E, Cantelli A. A model of submarine channel-levee evolution based on channel trajectories: Implications for stratigraphic architecture. Marine and Petroleum Geology. 2011;**28**:716-727

[33] Pickering KT, Hodgson DM, Platzman E, Clark JD, Stephens C. A new type of bedform produced by backfilling processes in a submarine channel, Late Miocene, Tabernas-Sorbas Basin, SE Spain. Journal of Sedimentary Research. 2001;**71**(5):692-704

[34] Fildani A, Hubbard SM, Covault JA, Maier KL, Romans BW,

Traer M, et al. Erosion at inception of deep-sea channels. Marine and Petroleum Geology. 2013;**41**(2013):48-61

[35] Covault JA, Sylvester Z, Hubbard SM, Jobe ZR. The Stratigraphic Record of Submarine-Channel Evolution. The Sedimentary Record. 2016;**14**(3):4-11

[36] McHargue T, Pyrcz MJ, Sullivan MD, Clark JD, Fildani A, Romans BW, et al. Architecture of Turbidite Channel Systems on the Continental. Marine and Petroleum Geology. 2011;**28**:728-743

[37] Tinterri R, Magalhaes PM, Tagliaferri A. Foredeep Turbidites of the Miocene Marnoso-Arenacea Formation (Northern Apennines). Milan: AAPG International Conference & Exhibition; 2012. p. 2011

[38] Hubbard SM, Ruig MJ, Graham SA. Confined channel–levee complex development in an elongate depo-center: Deep-water Tertiary strata of the Austrian Molasse basin. Marine and Petroleum Geology. 2009;**26**(41):85-112

[39] Hubbard SM, Covault JA, Fildani A, Romans BW. Sediment transfer and deposition in slope channels: Deciphering the record of enigmatic deep-sea processes from outcrop. GSA Bulletin. 2014, 2014;**126**(5/6):857-871

[40] Ricci ME, Lucchi FF. Turbidite facies and facies associations. In example of Turbidite Facies and Furies Associations from Selected Formations of the Northern Apennines. In: Mutti E, Parea GC, Ricci Lucchi F, Sagri M, Zanzucchi G, Ghibaudo G, Jaccarino S, editors. Nice: ZX International Congress Sedimentation; 1975. pp. 21-36

[41] Mutti E. Distinctive thin-bedded turbidite facies and related depositional environments in the Eocene Hecho Group (South-central Pyrenees, Spain). Sedimentology. 1977;**24**:107-131

[42] Rupk NA. Sedimentology of very thick calcarenite marlstone beds in a flysch succession, south-western Pyrenees. Sedimentology. 1976;**23**:43-65

[43] MacKenzie WS, Donaldson CH, Guilford C. Atlas of Igneous and Their Textures. England: Longman Scientific and Technical; 1982. p. 148

[44] Mohammad YO. Petrology of ultramafic and related rocks along Iraqi Zagros Thrust Zone. D.Sc thesis,. Osaka Prefecture University; 2008. p. 156

[45] Ismail SA, Arai S, Ahmed AH, Shimizu Y. Chromitite and peridotite from Rayat, northeastern Iraq, as fragments of a Tethyan ophiolite. Island Arc. 2009;**18**:175-183

[46] Yardley BWD, MacKenzie WS, Guilford CG. Atlas of Metamorphic Rocks and Their Textures. Longman Scientific and Technical, Longman Group UK Limited; 1990. p. 199

[47] Ajirlu MS, Moazzen M, Hajialioghli R. Tectonic evolution of the Zagros Orogen in the realm of the Neotethys between the Central Iran and Arabian Plates: An ophiolite perspective. Central European Geology. 2016;**59**(1-4):1-27

[48] Alavi M. Regional stratigraphy of the Zagros fold-thrust belt in Iran. American Journal of Science. 2004;**307**(9):1064-1095

[49] Aziz RM, Eliase M, Al-Hafdhn M. The igneous and metamorphic metaclastics of Naupordan Group and their relation to the tectonic regime in NE-Iraq. Mu'tah Journal Research and Studies. 1993;**8**:155-172

[50] Ali AA, Buckman S, Aswad KJ, Jones BG, Ismail SA, Nutman AP. Recognition of late cretaceous hasanbag ophiolite-arc rocks in the Kurdistan Region of the Iraqi Zagros suture zone: A missing link in the paleogeography of the closing Neotethys Ocean. Lithosphere. 2012;**4**(5):395-410

[51] Homke S, Verges J, Serra-Kiel J, Bernaola G, Sharp I, Garces M, et al. Late-cretaceous-paleocene formation of proto-Zagros foreland basin, Lorestan Province, SW Iran. Geological Society of America Bulletin. 2009;**121**(7-8):963-987

[52] Sarmad AA, Buckman S, Aswad KJ, Jones BG, Ismail SA, Nutmani AP. The tectonics evolution of a Neo-Tethyan (Eocene–Oligocene) island-arc (Walash and Naopurdan groups) in the Kurdistan region of the Northeast Iraqi Zagros Suture Zone. Island Arc. 2013;**22**:104-125

[53] Saura E, Verges J, Homke S, Blanc E, Serra-Kiel J, Bernaola G, et al. Basin architecture and growth folding of NW Zagros early foreland basin during Late Cretaceous and Early Tertiary. Journal of the Geological Society London. 2011;**168**:235-250

[54] Sarmad AA, Sleabi RS, Talabani MJA, Jones B. Provenance of the Walash-Naopurdan back-arc-arc clastic sequences in Iraqi Zagros Suture Zone. Journal of African Earth Sciences. 2016;**125**:73-87

[55] Sarmad AA, Ismail SA, Nutman AP, Bennett VC, Jones BG, Buckman S. The intra-oceanic Cretaceous (~108) Kata-Rash arc fragment in the Kurdistan segment of Iraqi Zagros suture zone: Implication for Neotethys evolution and closer. Lithos. 2016;**260**:154-163

[56] Middleton GV, Hampton MA. Subaqueous sediment transport and deposition by sediment gravity flows. In: Stanley DJ, Swift DJP, editors. Marine Sediment Transport and Environmental Management. New York: John Wiley and Sons; 1976. pp. 197-218

[57] Calvo C. Provenance of plutonic detritus in cover sandstones of Nicoya Complex, Costa Rica: Cretaceous unroofing history of a Mesozoic ophiolite sequence. GSA Bulletin. 2003;**115**(7):832-844

Chapter 3

Analyzing Sedimentary Rocks to Evaluate Paleo Dimensions and Flow Dynamics of Permian Barakar River of Rajmahal Gondwana Sub-Basin of Eastern India

Zahid A. Khan and Ram Chandra Tewari

Abstract

The cross strata and planar strata of sedimentary rocks are used in estimating paleo dimensions and flow dynamics of Permian Barakar River of eastern India. Quantitative estimates indicate that mean bed form became thinner *i.e.,* from 1.363 m to 0.928 m to 34.21 m–23.33 m, as the Barakar River flowed with flow velocity of between 90 cm/sec and 157 cm/sec from Fox Ridge situated to the south on a steeper slope which becomes gentler downstream. The friction factor (0.036) is on higher side indicating low flow resistance while bed shear stresses remain competent, even during low paleo-discharges to transport coarse-pebble bed. Rouse number, Z, is decreased from Brahmini (3.38) to Hurra (1.71) implying predominant bed-load in upstream and local transition to mixed load in the downstream. Bed load formed 17.74 to 15.62 percent in upstream and reduced to 10.14–7.94 percent in downstream suggesting bed load channel for the former and mixed load channel for the latter corroborating estimated Rouse Number (Z) values. The trunk Barakar River on an average was about 2260 km long, 817 m wide and 13.63 m deep with channel belt width in between 6310 and 2205 m. The river channel sinuosity was in place of sinuosity 1.361 in the south-southwest part and progressively became more sinuous in the northern and northeastern part of the basin (1.728) with maximum sinuosity of 1.955. The catchment area of the Barakar River lies between 10,700 to 422,600 km^2 and paleo-discharge between 22,070–4510 m^3/sec with the maximum 66,000 m^3/sec and whose mean annual flood was in the range of 170,600–42,260 m^3/sec. These parameters suggest that initially multi-thread and broad Barakar River became narrow and single thread in downstream and had its outlet in the northeast (Sikkim, Assam and Bangladesh) towards the Tethys Ocean during Permian times.

Keywords: sedimentary rocks, paleo dimension, flow dynamics, Permian, Barakar, and Gondwana

1. Introduction

Internal sedimentary bed forms such as dunes and ripples, which are believed to be truncated in the preserved geologic record, can now have their full heights estimated from cross bedding-sets and cosets within quantitative uncertainty and order of magnitude [1–4]. Recent researches provide a methodology to evaluate paleo dimensions and flow dynamics of ancient fluvial systems [5–7]. Fluvial paleo-channels can be scaled from numerical equations based on grain size along with channel depth and width measurements and augmented by flow depth estimated from estimated dune bed form height from cross bedding set thickness using data of cross bedding set thickness [2, 3, 8]. Calculated parameters include mean bed form height, channel depth, channel belt width, channel width, paleoslope, boundary shear stress; Darcy-Weisbach friction factor, paleoflow velocity, and paleo discharge provide insight into basin analysis.

An attempt is made to scale the paleo-dimensions and flow dynamics of fluvial channels of Early Permian Barakar River. These parameters together with sedimentary rock record provide meaningful interpretations of fluvial architecture, the evolution of fluvial style.

2. Study area, stratigraphy and sedimentological sketch

The Rajmahal sub- basin of eastern India is part of the Gondwana Master basin and covers a large area of Bengal basin, North Bengal and Purnea. There are four sub basins (coalfields) from south to north i.e. Brahmini, Pachwara, Chuperbhita and Hurra (**Figure 1A**). The Gondwana rocks are bounded by the river Ganga to the north, by Rajmahal volcanic to the east, by rugged metamorphic to the west and by laterite and recent alluvium to the south. The unclassified Archean rocks form the basement of the Gondwana rocks in the Rajmahal sub-basin with a pronounced unconformity, though at places the contact is faulted. Isolated patches of Talchir rocks are seen along the western periphery of the basin. This formation is succeeded by the Karharbari Formation, composed of grits, conglomerates and sandstones. The overlying coal bearing Barakar Formation directly rests over the Archean basement at most places and is exposed along the river valleys where the overlying traps and Triassic sediments are denuded. The Early Permian Barakar rocks are overlain by the Dubrajpur Formation (Late Triassic), which is, intern, followed by Rajmahal traps and Inter-trappean rocks (Early Cretaceous to Early Jurassic).The contacts between Early Permian, Late Triassic and Early Cretaceous are unconformable. A generalized stratigraphic sequence of the Gondwana rocks of Rajmahal sub-basin is shown in **Figure 1B**.

Gondwana sedimentation in Rajmahal sub-basin commenced with the deposition of the Talchir sediments under glacial and fluvio-glacial, glacio-lacustrine and/or shallow marine environments [9]. Following retreat of Talchir ice, the rejuvenated rivers flowed dominantly towards the northwest and northeast and deposited shallow waters gravelly and sandy sediments as discontinuous patches gradationally above Talchir as Karharbari Formation. The overlying Barakar rocks are characterized by fining upward cycles of coarse to medium sandstones with interbedded siltstone, shale and thick to thin coal seams [10, 11]. These rocks were deposited by laterally migrating river channels in response to varying discharge and/or intermittent differential subsidence/tectonism, and led to the development of peat swamps in the flood plains and protected lakes.

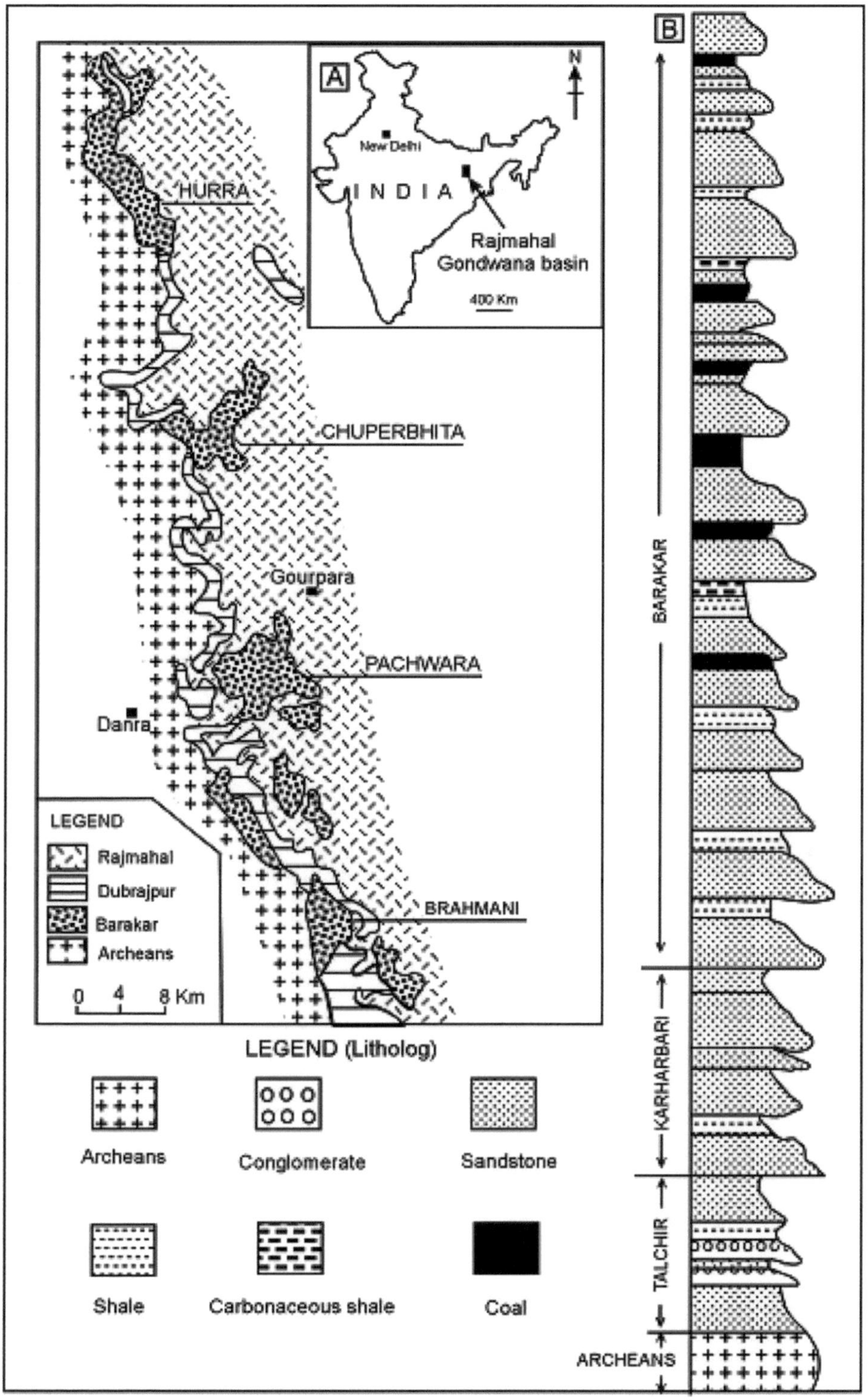

Figure 1.
(A) Location and geological map; and (B) subsurface Gondwana stratigraphy of Rajmahal sub-basin of eastern India (based on bore hole log).

3. Methodology

3.1 Field observations and basic data

Sedimentological data is collected from 26 localities measuring 5–10 meter vertical sections mostly along the river and small tributaries in all the four areas. Once all outcrops were located, multiple measurements were collected including set-thickness, average grain sizes, and total thickness. Cross-bed sets thickness were measured in the z-plane with 30 cm scale and the minimum distance between measured vertical sections was roughly 0.50 km. The cross bedding sets in the study area are interpreted to be mostly dunes for the following reasons:

(1) cross bedding sets are truncated, (2) Paleocurrent directions were much less variable than one would expect in bars, (3) even if there are some bars within the data, they are usually made up of many layers of truncated dunes, and (4) all ripples observed on the outcrops were flowing in the same direction as the cross bedding sets rather than in a separate direction as see on a bar [12, 13].

3.1.1 Cross-sets

Cross-set heights were measured as these data can be used to reconstruct original bedform heights and formative flow depths. Trough- and planar cross bedding, which are genetically indicative of bed load transport, were present at nearly all studied field sites. They occurred predominantly in sand grade deposits, but omnipresent in pebble grade deposits of southern coalfields. To establish mean cross-set heights, the sampling strategy of [7] was followed.

3.1.2 Grain size

Grain sizes were measured at each outcrop. These measurements are taken using a standard 10× hand lens and a grain size card. Grains were identified in each set according to Udden-Wentword grain size scale. Grains within each set tended to be unimodal or at least largely represented by one size; this made finding the average size within cross bedding sets and was substituted for D_{50} which represents the median of the grain size distribution used in this study.

Sedimentological data so acquired from the Barakar exposures include grain size, cross bedding height, and bar-form height. These data were then used to determine multiple channel geometry, paleohydraulic parameters and paleo dynamics including: mean bed form height, channel depth and width, channel belt width, paleoslope, boundary shear stress, Darcy-Weisbach friction factor, paleoflow velocity, paleodrainage, and overall drainage area by the methods outlined below.

4. Quantitative paleo-dimensions

4.1 Bedform height and channel depth (H_m) & (Dc)

To calculate original bedform heights from average cross-set measurement, the empirical relationship [14] was used. These equations determined relationships between the mean value of the exponential tail of the probability density function

(PDF) for cross-set thicknesses and dune heights [14, 15]. The work [14, 16] has shown that the dune height (H_m) can be estimated from mean cross–set thickness (S_m) using a regression equation:

$$H_m = 5.3\,\beta + 0.001\,\beta^2$$

Where $\beta = S_m/1.8$.
This can be simplified as:

$$H_m = 2.90\,(\pm 0.70)^*\,S_m \tag{1}$$

Where 2.90 are the mean and 0.70 is the standard deviation. H_m is the mean dune height; S_m is the mean cross-set thickness, β is the mean value of the exponential tail of the probability density function (PDF) for topographic height relative datum. While Eq. (1) relating formative dune heights to preserved cross bedding sets perhaps unusual in a wide variety of natural forces and random events accompanying the formation of such cross-sets. Nevertheless' the above equation is valid for a wide range of flume and river experiments [1, 2] to suggest that abnormally thick, isolated cross bedding sets formed by unit bars should be excluded from the data. The formative dune height is calculated from the preserved cross-sets and erosional surfaces at the top and bottom of cross bedding sets are almost horizontal so it is unlikely that aggradations and degradations played any role in the expression of cross-sets within the Barakar sandstone.

Equations for relating dune height to channel depth were developed [14] (Eq. (2)) show that flow depth is typically 8 to 10 times mean dune height. Alternatively, ref. [17] empirical relationship (Eq. (3)) based on the relation between bankfull depth and dune height referred to as $Dc = 11.6\,H_m^{0.84}$ can be used as a measure of maximum channel bankfull flow depth (Dc)

$$8 < Dc/H_m < 10 \tag{2}$$

$$Dc = 11.6\,H_m^{0.84}, 0.1\text{ m} < Dc < 100\text{ m} \tag{3}$$

Where Dc is river channel depth and H_m is mean dune height.

The method of using cross-beds to determine channel flow depth has been tested in several modern and ancient rivers [14, 15, 18]. Based on measured sections thickness of dune scale cross-strata sets are on average 47 ± 31 cm (Brahmini sub-basin); 43 ± 28 cm (Pachwara sub-basin); 37 ± 20 cm (Chuperbhita sub-basin); and 31 ± 14 cm (Hurra sub-basin)' respectively. Setting (S_m) = 47—30 cm, the above Eq. (1) yields a mean dune height (H_m), varying from 0.928–1.363 m for the above four Gondwana sub- basins from south to north. These reconstructed mean dune heights agree closely with those reported by ref. [2] of 0.90–1.52 m and ref. [1] of 1.07–1.42 m from ancient fluvial channel bars. According to ref. [14] bankfull flow depth is typically 8 to 10 times mean dune height. This suggests bankfull flow depths (Dc) of between 7.25–13.63 m. Using ref. [17] relationship, average bankfull channel depth of Barakar sandstone is varying between 10.87–14.96 m. We thus determine maximum (i.e. bankfull) flow depths to be about 13.0 m.

The paleochannel flow depth was estimated from the thickness of lateral macroform using the equation [19]; a 10% reduction in thickness for the conversion of sand to sandstone as a result of compaction appears conservative following Ingles and Grant suggestion, therefore, divided by 0.9 to convert these measures to original bankfull depth.

$$Dc = D^*/0.9 \tag{4}$$

Where D^* is the maximum channel bankfull flow depth, which is represented by the thickness of the sandstone macroform, 0.9 compensates for the compaction factor.

The average thickness of the fining-upward facies sequence of the Barakar sandstone varies from 15 m in Brahmini sub-basin in the north through Pachwara sub-basin and Chuperbhita sub-basin to 10 m in Hurra sub-basin in south. The bankfull depth so calculated yield a value in between 11.12–16.67 m which is similar to those calculated from Eqs. (2) and (3) calculated above.

4.2 Bedform length (L_b)

Ref. [20] have presented a model relating bedform length (L_b) to bedform height (H_m) as:

$$H_m = \partial_0 L_b{}^{\partial}{}_1$$

Where ∂_0 and ∂_1 are empirical parameters with associated uncertainties and values are $\partial_0 = 0.0513$; $\partial_1 = 0.7744$ [21] (**Table 1**) and L_b is bedform length. Substituting the values of empirical parameter in Eq. (5) we get

$$H_m = 0.0513\, L_b{}^{0.7744} \tag{5}$$

An empirical relationship between bedform length and bed forms height has been presented by ref. [22] as:

$$H_m = 0.0677\, L_b{}^{0.8090} \tag{6}$$

On the other hand, ref. [23] who used the data from the Rio Parana (Argentina) and those from the Lower Rhine (The Netherland) and Negro (Brazil) obtained relationship between bed form height and length which affords another assessment of the bed form length.

$$H_m = 0.21\, L_b{}^{0.560} \tag{7}$$

When previously estimated value of H_m is substituted, Eq. (5) indicates that the bedform length of dunes was about 34.21 m in Brahmini sub-basin and decreased down the paleoslope to about 23.33 m in Hurra sub-basin; Eq. (6) indicates bedform height in between 25.52 m and 17.65 m, and from Eq. (7) it is 28.18 m and 14.16 m in up dip basins (Brahmini, Pachwara) and decreased down dip to 18.36 m and 14.16 m in down dip basins (Chuperbhita, Hurra). The overlap in these three approximations suggests that the bedform length of the Barakar River was about 34.21 to 14.16 m. These dunes are classified as large dunes and such dunes become progressively smaller and their lengths become shorter as the flow strength decreases. For equilibrium dunes, the maximum steepness, H_m/L_b. is approximately 0.055 (L_b = dune wavelength), L_b/D_c is approximately 6, and the minimum D_c/H_m is approximately 3. It appears that for all types of river dunes irrespective of those not in equilibrium with the flow in the Barakar sandstone have Dc/H_m average between 6 and 10. Recently ref. [24] suggests that steeper slopes tend to develop elongate, narrow braid bars than rivers with shallower slopes supports above contention that during deposition of Lower Barakar sediments, the River had long braid bars ($\approx$35 m) than during deposition of Upper Barakar ($\approx$15 m).

Parameter	Empirical equation	Brahmini sub basin	Pachwara sub basin	Chuperbhita sub basin	Hurra sub basin
Mean cross-bed thickness (*m*)	S_m	0.47	0.43	0.37	0.31
Mean particle Size (*mm*)	(D_{50})	0.72	0.57	0.45	0.35
Dune height (H_m)	$H_m = 2.90\ (\pm 0.70)\ S_m$	1.363	1.247	1.073	0.928
Flow Depth(*m*)	$8 < Dc/H_m < 10$	10.91–13.63	9.64–12.52	8.58–10.73	7.25–9.23
	$Dc = 11.6\ H_m^{0.84}$	14.96	13.87	12.56	10.87
	$Dc = D^*/0.9$	13.54	12.48	11.30	9.78
Bedform-length (*m*)	$H_m = 0.0513 L_b^{0.7744}$	34.21	31.34	26.97	23.33
	$H_m = 0.0677 L_b^{0.8090}$	25.52	23.27	20.77	17.65
	$H_m = 0.21 L_b^{0.56}$	28.18	23.99	18.36	14.16
Channel width (*m*)	$W_c = 8.88 * D_c^{1.82}$	817	697	534	408
	$Wc = 64.4 * D_c^{1.54}$	2956	2586	2063	1380
Channel belt width (*m*)	$Wcb = 59.9\ D_c^{1.80}$	6310	5534	4305	2205
	$W_c = (Wcb/6.89)^{1.01}$	6780	5370	3240	2800
Sediment load parameter (*M*)	$F = 255\ M^{-1.08}$	3.07	3.52	5.41	6.92
Percentage of total load as bed load	55/M	17.74 *Bed load channel*	15.60 *Bed load channel*	10.14 *Mixed load channel*	7.94 *Mixed load channel*
Rouse Number (*Z*)	$Z = W_s/\beta\kappa\ U^*$	3.58	3.46	2.01	1.71
Reynolds Particle Number (Re_p)	$Re_p = \sqrt{R\ gD_{50}} D_{50}/\upsilon$	79.957	49.415	29.573	16.975
Channel Slope (S_c)	$Sc = \tau^*_{bf50} R D_{50}/D_c$	0.000162	0.000152	0.000139	0.000116
	$\log S_c = \alpha_0 + \alpha_1 \log D_{50} + \alpha_2 \log D_c$	0.000138	0.000144	0.000159	0.000176
	$Sc^{½} = n V_c/D_c$	0.000065	0.000052	0.000045	0.000038
Froude Number (F_r)	$F_r = V_c/\sqrt{g\ D_c}$	0.136	0.111	0.108	0.102
Boundary stress shear *N/m² or Pa*	$\tau_b = \rho\ g\ D_c S_c$	10.68	8.66	6.77	5.24
Critical shear stress *N/m² or Pa*	$\tau_{cr} = \tau^*\ (\rho_s - \rho_w)\ g\ D_{50}$	0.169	0.152	0.174	0.204
Velocity(*m/sec*)	$V_c = (R^{0.67} Sc^{0.50})/n$	1.57	1.35	1.12	0.92
	$V_c = [(8\ g\ R\ Sc/f)]^{0.50}$	1.89	1.77	1.52	1.27
Manning Roughness Coefficient (*n*)	$n = 0.039\ D_{50}^{0.167}$	0.036	0.025	0.023	0.021
Darcy-Weisbach Friction factor (*f*)	$(8/f)^{0.50} = 2.2\ (D_c/D_{50})^{-0.055} Sc^{-0.275}$	0.037	0.036	0.035	0.033
Channel sinuosity (S_{ch})	$S_{ch} = 3.5 (W_c/D_c)^{0.20}$	1.361	1.517	1.621	1.632

Parameter	Empirical equation	Brahmini sub basin	Pachwara sub basin	Chuperbhita sub basin	Hurra sub basin
	$\text{Log } S_{ch} = 3.68 - 0.0684 \text{ L} + 0.00032 \text{ L}^2$	1.383	1.570	1.663	1.728
Discharge *(m^3/sec)*	$Q_w = W_c D_c V_c$	22,070	12,865	5880	4510
Mean Annual flood *(m^3/sec)*	$\text{Log } Q_{flood} = 2.084 - 0.070 \log D_A{}^2 + 0.865 \text{ Log } D_A$	170,600	112,200	65,450	48,640
Drainage Area *(km^2)*	$D_A = 249\ D_c{}^{2.44}$	107,004	86,536	60,512	42,260
	$Q_w = 0.0161\ D_A{}^{0.9839}$	136×10^3	81×10^3	37×10^3	28×10^3
Length of River *(km)*	$L = 1.109\ D_A{}^{0.545}$	703	631	525	427

Table 1.
Paleo- dimensions, flow dynamics and paleochannel parameters of early Permian Barakar River.

4.3 Channel width (W_c)

There are a number of empirical equations relating average bankfull channel depth to river channel width, many of which are thoroughly communicated and evaluated [13, 25]. However, the empirical equations used here are those they developed on river depth to channel width [1, 26] as

$$W_c = 8.88^*\ D_c{}^{1.82} \tag{8}$$

$$W\text{ c} = 64.4^*D_c{}^{1.54} \tag{9}$$

Where, W_c = width of channel, and D_c = bankfull channel depth. Put down the value of channel depth in Eq. (8) we get the width of channel in between 408 m −817 m, and Eq. (9) indicate about 1380 m to 2956 m in the Rajmahal Barakar River.

4.4 Sediment load parameter (F)

So important is the type of sediment load in determining fluvial morphology that Schumm has proposed classifying alluvial river channels on the basis of the ratio of suspended load to bed-load and introduced the term bed-load, mixed load and suspended load channels ([27], p. 1579, **Table 1**). Although the ratio of suspended load to bed load is not determined directly but directly related to the percentage of silt and clay in the channel perimeter. And to calculate the percent silt-clay in the channel perimeter (M), following relationship is used:

$$\text{M} = (\text{S}_\text{a}{}^*\ W_c) + (\text{S}_\text{b}{}^*\ D_c)/W_c + D_c \tag{10}$$

Where M = Sediment load parameter, W_c = Bankfull channel width, D_c = Bankfull channel depth, S_a = percent silt-clay in channel alluvium, and S_b = percent silt-clay in the bank alluvium. Detailed petrographic and geochemical studies in recent years have revealed that much of the clay size fraction of Barakar sandstones is authigenetic and secondly digenesis may completely alter the original silt-clay matrix may cause unknown error in estimation of M derived from Eq. (10), hence not used in present study. It is also possible to estimate M for the Barakar sandstone indirectly from Schumm's following equation:

$$F = 255\ M^{-1.08} \tag{11}$$

Where F (width/depth ratio) = $W\,c/Dc$, and W_c = bankfull channel width, Dc = bankfull channel depth, and M = sediment load parameter. Substituting the estimated values of Wc and Dc, Eq. (11) indicates that Barakar River had a sediment load parameter of 3.07 and 3.52 for the Brahmini and Pachwara sub basin and between 5.41–6.92 in the northern sub basin of Chuperbhita and Hurra. Values from tables ([27], p. 1579, **Table 1**) indicate range in M, between 5 and 20 for mixed load rivers and less than 5 for bed load Rivers. Furthermore, the assessment of percentage of total load as bed load is estimated by using ref. [27] formula:

$$55/M$$

where M denotes sediment load parameter.

By substituting values for M, the bed load comes out to be 17.74, 15.62, 10.14 and 7.94 respectively for Brahmini, Pachwara, Chuperbhita and Hurra sub basin respectively. Calculated values of M and percentage of total load as bed load in this study fall within the range of those established values for bed load rivers in the southern part (Brahmini and Pachwara sub basins) and mixed load rivers in northern part (Chuperbhita and Hurra sub basin).

4.5 Width of channel belt (W_{cb})

There are a number of empirical relationships between channel width and channel belt width, for example, from ref. [25] (Eq. (12)) and [1] (Eq. (13)).

$$W_{cb} = 59.9\ D_c^{1.80} \tag{12}$$

$$W_c = (W_{cb}/6.89)^{1.01} \tag{13}$$

Taking the mean bankfull channel depth of 15.00 m–11.78 m gives a range of channel-belt widths from 6780 m–2205 m. The value in these calculations is that they give a range of possible channel and channel-belt sizes that compare favorably with outcrop observations where the Barakar channels are confined, or where channel margin are observable.

Sedimentological descriptions of the fluvial deposits of the study area show a predominance of dune-scale cross-stratified, pebbly, coarse to medium grained sandstone, fining upward into fine grained sandstone and shale, shaly coal and thick coal seams characterized by asymmetrical fining upward cycles [11]. The abundant shale and fine clastics deposited in the floodplain demonstrate that the Barakar Rivers carried a significant shaly load in suspension, as well as a sandy bed load moving primarily as dune-scale bedform.

4.6 Paleo-channel slope (Sc)

Slope affects river plan form (i.e. braiding is favored by steeper slope and meandering by gentler slope) and facies boundaries [13, 28]. Paleoslope in ancient river system is calculated using physics-based methods, such as Shield's [29] threshold for bed-load movement, especially where grain size, channel depth and channel width of a paleo-river are known [4, 16]. Paleoslope is estimated using grain size and density of sediment grains following the empirical equation [22, 29];

$$\tau^*{}_{bf50} = (D_c Sc)/(RD_{50})$$

on re-arranging we get

$$Sc = \tau^*{}_{bf50} RD_{50}/D_c \quad (14)$$

Where *Sc* is the paleoslope, τ^*_{bf50} is the bankfull Shields number for dimensionless shear stress, *Dc* is the mean bankfull channel flow depth, R is the submerged dimensionless density of sand-gravel sediment ($\rho_{s-}\rho_w$), ρ_s = the grain density (assumed to be quartz, with density of 2.65 g/cm^3), ρ_w = the fluid density; and D_{50} is the median grain size, τ^*_{bf50} is assumed to be 1.86 [22]. When working with ancient fluvial systems, D_c is measureable from outcrop or estimated by empirical formula (Eqs. (7) and (8)), and D_{50} for this study was estimated from observations of rocks in outcrop using magnifying lens 10× and/or from grain size analysis. When previously determined values of channel depth and median grain size are substituted, Eq. (13) indicate that paleochannel slope or gradient was about 0.000162 for Brahmini sub basin and decreased to 0.000116 in northern Hurra sub basin as sedimentation progressed in the area.

Another estimate of the paleoslope of the Early Permian Barakar River can be derived using the model [23] of which is based on Bayesian regression analysis relating paleoslope to bankfull channel depth, *Dc*, and median grain size D_{50} that takes the form:

$$\text{Log } S_c = \alpha_0 + \alpha_1 \log D_{50} + \alpha_2 \log D_c \quad (15)$$

Where $\alpha_{0,}$ α_1, and α_2 are empirical constant parameters with associated uncertainties are given by −02.08 for α_0, 0.254 for α_1, and − 1.09 for α_2 respectively [30] (**Table 1**). Empirical paleoslope reconstruction, such as the one applied here for the Barakar rivers; typically involve approximately an order-of-magnitude variation fitted to Trampush's model. Ultimately, this may have a significant control on the uncertainty of any quantitative estimation of past flow conditions. The sediment grain sizes of the Brahmini and Pachwara channel rages from very coarse to coarse grained sand (≈ 0.72–0.57 mm) with the fine grained sand (≈ 0.25 mm). The grain sizes of remaining two sub basins (Chuperbhita and Hurra) yielded a D_{50} grain size value between 0.45 to 0.35 mm, categorized as medium to fine grained sandstone. By substituting values for D_{50} (0.72, 0.57, 0.45 and 0.35 mm), D_c (13.63–9.28 m) in Eq. (15), the channel paleoslope is calculated to between 0.000176 and 0.000138.

Additional approximation of the paleoslope can be derived using the Manning equation for the open channel flow systems. This same relationship has been used to determine the paleochannel slope in many ancient fluvial rivers [5, 8].

$$V_c = 1.486 \left(R^{2/3} S_c{}^{1/2}\right)/n \quad (16)$$

In this equation, V_c is the average flow velocity, R is the hydraulic radius, approximately equal to the channel depth, D_c, S_c is the channel paleoslope, and *n* is the Manning roughness coefficient (For metric calculations (CGS), 1.486 is replaced by 1 because Manning equation employed FPS system). Rearranging Eq. (16) and putting $R = D_c$ and 1 in place of 1.486, the formula becomes:

$$Sc^{1/2} = nV_c/D_c \tag{17}$$

Paleoslope of the Barakar channels estimated using Eq. (17) ranges from 0.00038 to 0.00065 (**Table 1**), which suggests a low paleoslope comparable to slope ranges for the Amazon, Mississippi and Niger Rivers (slope range 0.00002–0.0005; 21). The estimated moderately high paleoslope from Eqs. (15)–(17) for the Brahmini and Pachwara sub basins are an order of magnitude equivalent to the estimated slope of the low sinuosity channel, whereas, estimated lower slope for Chuperbhita and Hurra sub basin channels agrees with the dominant fine clastics observed in the sand body. These values are within the low end of the range of values recorded from modern, humid-tropical climate, fluvial systems that transported mixed load sediments.

5. Flow dynamics

5.1 Boundary shear stress (τ_b)

The boundary shear stress (τ_b) acting on the bed of the channel can be calculated as:

$$\tau_b = \rho\, g\, D_c\, S_c \tag{18}$$

where τ_b = the boundary shear stress, ρ = the fluid density, g = gravitational acceleration, D_c = averaged channel flow depth, and S_c is averaged water-surface paleoslope. Both field and laboratory experiments have shown that initial motion of bed materials in coarse-medium grained rivers typically occurs at a transport stage between 1 and 3 [31]. This relationship between the flow and its container can be applied to all natural channels with some error and has been recently applied in ancient fluvial deposits [32] of the Sharon Formation, USA, ref. [33] for the Parthenon Sandstone USA. Using previous estimates of Dc and Sc in Eq. (16) the boundary shear stress is estimated to be in between 10.68–5.24 N/m^2 for the Barakar Rivers in Rajmahal Gondwana master basin.

The critical shear stress (τ_{cr}) represents the necessary boundary shear to move the bed-load materials, based upon their grain size, grain shape, effective density, and roughness. Therefore, the formulation to express critical shear stress (τ_{cr}) for non-cohesive sand is provided by Shield [29] and is given as:

$$\tau_{cr} = \tau^* \ (\rho_s - \rho_w)\, g\, D_{50} \tag{19}$$

Where τ_{cr} = the critical shear stress; τ^* = the Shield number for the given particle non-dimensional critical shear stress; ρ_s = the grain density (assumed to be quartz with a density of 2650 kg/cm^3; ρ_w = fluid density 1000 kg/m^3; g = the acceleration due to gravity in m/sec^2 and D_{50} = median particle size in meter. Critical shear stress is calculated from solving Eq. (19) when the transport stage was set to initial motion i.e., 1, using paleohydrological data for bankfull depth, and grain size data which comes out in between 0.169–0.204 N/m^2 or Pa.

Sediment mobility for a given particle size occurs when the boundary shear stress exceed the critical shear stress, in other words $\tau_b > \tau_{cr}$. This relationship has been observed in the Barakar sandstones of present study.

5.2 Paleoflow velocity (V_c)

Under the assumption of steady, uniform (i.e. constant channel depth) flow in a channel that has a rigid boundary (i.e. flow conditions up to bed motion), the threshold mean velocity, V_t, can be computed based on resistance coefficients and channel geometry by two commonly and widely used methods. The first employs the Manning roughness coefficient, *n*:

$$V_c = \left(R^{0.67} Sc^{0.50}\right)/n \tag{20}$$

And the second uses the Darcy-Weisbach friction factor, *f*:

$$V_c = [(8\,g\,R\,Sc/f)]^{0.50} \tag{21}$$

Where R is the hydraulic radius is approximately equal to the depth D_c, Sc is the channel slope and *n* is the Manning roughness coefficient. The threshold mean flow velocity is a function of the size of bed particles that have been on the way, as evident in the equation for determining the resistance coefficient *n* and *f* below [Eqs. (21) and (22)].

Unlike the Manning empirical equation, the Darcy-Weisbach equation uses a dimensionless friction factor, has a sound theoretical basis, and exact accounts for the acceleration from gravity; moreover, the relative bed roughness does not influence the exponents of hydraulic radius and channel slope. For these reasons, the Darcy-Weisbach equation is preferred over the Manning approach as discussed [34]. Many algebraic manipulations has been used to calculate Manning roughness (*n*) and Darcy-Weisbach friction factor (*f*) and then various relationships have been proposed between resistance coefficients (*n*, *f*) and sediment grain size [35] as:

$$n = 0.039\, {D_{50}}^{0.167} \tag{22}$$

Ref. [34] derived a roughness coefficient encompassed a range of straight, braided and meandering, bed sediment sizes (sand and gravel), and river sizes (small scale to large river), as well as perennial rivers:

$$(8/f)^{0.50} = 2.2\,(D_c/D_{50})^{-0.055} Sc^{-0.275} \tag{23}$$

Paleoflow velocities in the Barakar Rivers can be estimated using the method [36]. Their method is based on specific bed forms (e.g. ripples and dunes) are stable within specific ranges of grain size, flow velocity (**Figure 2**).

According to ref. [14] the height of a dune is strongly dependent on flow depth, therefore dune-scale cross sets is particularly useful in determining both flow velocity and water depth. If above assertion is correct than a flow velocity of between 65 and 160 cm/sec is estimated based on grain size (fine sand to coarse sand) using bed forms, grain size, and water depth and flow velocity plot [36]. Using this Eq. (23), the roughness coefficient (f) comes out to be between 0.0372–0.0334 (**Table 1**).

5.3 Rouse number (Z)

To determine dominant mode of sediment transport, the non-dimensional scale parameter Rouse number, Z, was calculated as

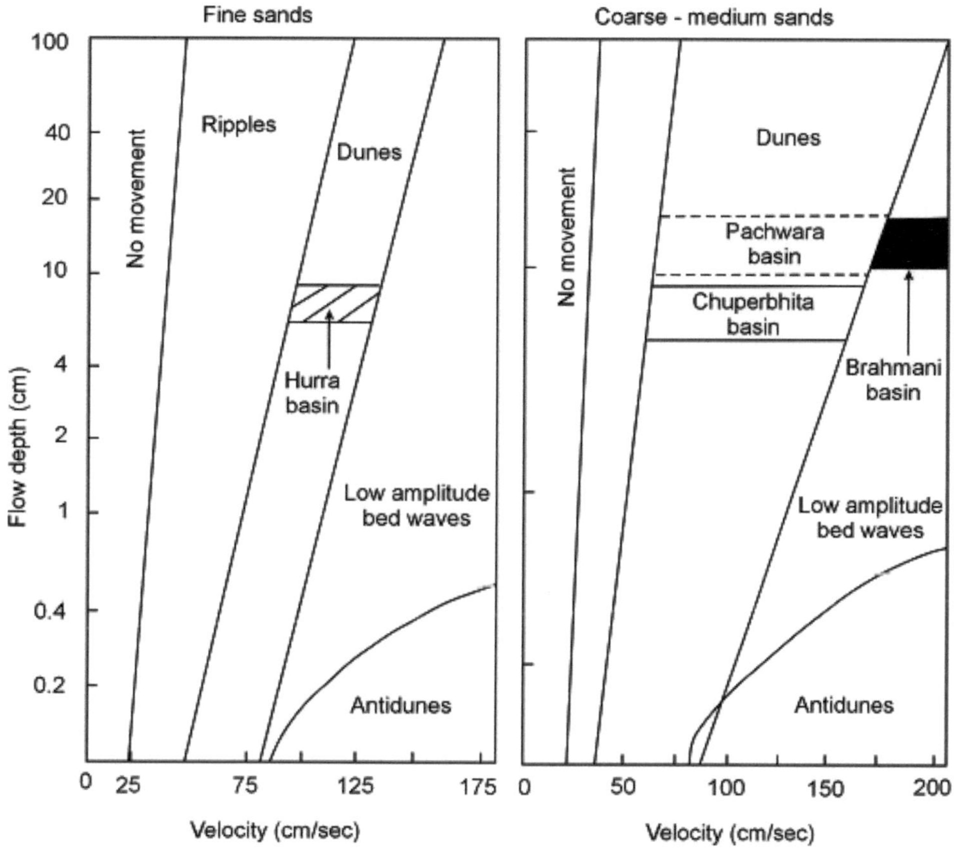

Figure 2.
Fine and coarse-medium grained bedform phase diagrams [37]. Inferred range of velocity for Hurra coal basin is 80–125 cm/sec. Pachwara coal basin is 60–150 cm/sec. Chuperbhita coal basin 58–160 cm/sec. Brahmini coal basin 150–200 cm/sec.

$$Z = W_s/\beta\kappa\, U_* \quad (24)$$

Where β is a constant taken as 1, and κ is the von Karman constant, taken as 0.40, U_* is the boundary shear velocity, and sediment settling velocity, W_s, was calculated as a function of grain size following [38] as;

$$Ws = R\,g\,D_{50}^{\ 2}/C_1\,v + \left(0.75\,C_2\,R\,g\,D_{50}^{\ 3}\right)^2 \quad (25)$$

Where g is the Earth's gravitational acceleration, D_{50} is the median diameter of a particle, v is the kinematic viscosity of water (1×10^{-6} for water at 20° C and $C_1 = 18$ and $C_2 = 1$ are constants associated with grain sphericity and roundness. And boundary shear velocity U_* determined as.

$$U_* = \sqrt{\tau_b}/\rho_w$$

Where τ_b is the boundary shear of the fluid and ρ_w is the mass density of the fluid. With Z, dominant mode of sediment transport in alluvial system is typically bed load for $Z > 2.5$, 50% suspended load (i.e. mixed load) for $1.2 < Z < 2.5$. Setting previously estimated values of R, Ws (sediment fall velocity), D_{50}, and kinematic viscosity of water, Eq. (24) indicates that the Rouse Number (Z) was about 3.58 and 3.46 for the Brahmini and Pachwara and decreased to 2.01 and 1.71 in the northern Chuperbhita and Hurra sub-basins. These estimated values for Z is characteristic of a bed load channel in the southern part and mixed load (50% suspended load) for the northern part of the Rajmahal Gondwana master basin. These estimated values are in agreement with Schumm sediment load parameter (M) calculated by Eq. (11) suggests convincingly that the southern part of the basin predominantly deposited by bed load channels whereas the northern part mostly deposited by mixed load channel.

5.4 Reynolds particle number (R_{ep})

To collaborate inferred sediment transport modes, the particle Reynolds particle number, R_{ep}, was additionally calculated as:

$$Re_{\mathrm{p}} = \sqrt{R\, g\, D_{50}} D_{50}/v$$

And plotted R_{ep} as a function of τ^*, following [39] (**Figure 3**). This plot enable field results to be contrasted with data that are typical of either bed load, mixed load and suspended load sediments [39], and to identify where these data are positioned among characteristics flow regimes (no sediment transport, ripples and dunes, upper plane beds) following ref. [40]. Using the statistical package **VCALC** the value of Reynolds number (R_{ep}) comes out to be 79.957 and 49.415 for the coalfields of southern part and decreasing to 29.573 to 16.975 in the coalfields of northern part of Rajmahal Gondwana master basin

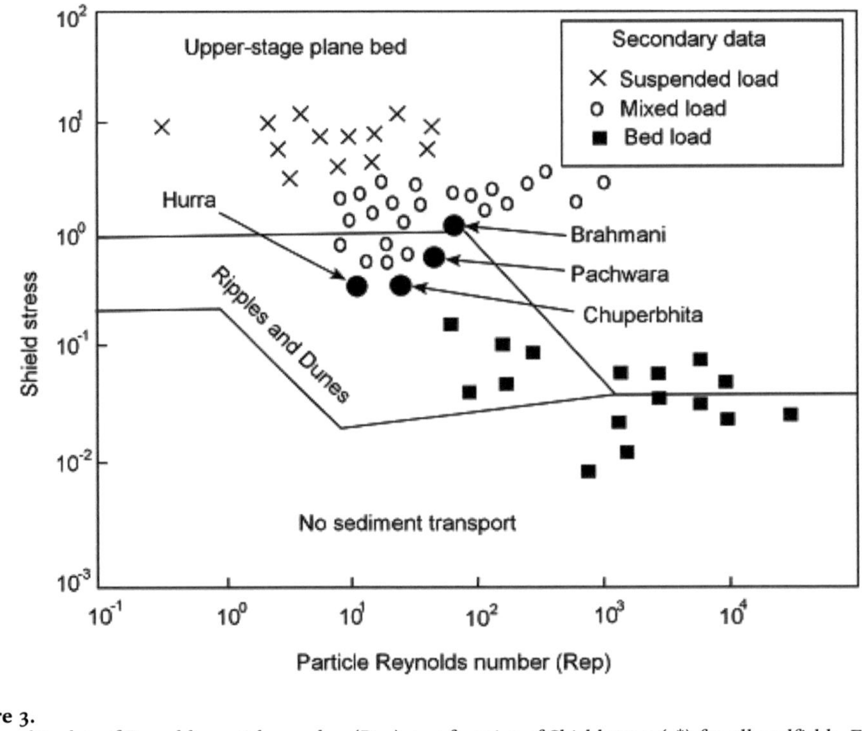

Figure 3.
The graphic plots of Reynolds particle number (Rep) as a function of Shield stress (τ^*) for all coalfields. For comparison, this plot includes secondary data originally compiled by ref. [39].

and corresponding sediment fall velocity in the range 0.1110–0.0485 m/sec. When these calculated values are plotted in **Figure 3** indicate that mostly Barakar sediments were deposited in dunes and ripple flow attributes corroborating outcrop field study.

5.5 Froude number (F_r)

The validity of paleoflow bankfull depth and paleo-velocity estimates was assessed by ensuring that the Froude number, Fr, mathematically written as:

$$F_r = V_c/\sqrt{g}\,D_c \tag{26}$$

Where Vc = water flow velocity, D_c = Bankfull channel depth, and g = acceleration due to gravity (approximately half of the present during Permian times, i.e. 4.9 m/ sec^2. The Froude number is a dimensionless parameter that describes different flow regimen of an open channel flow and is a ratio of inertial and gravitational forces. In open channel flow, the nature of the flow (supercritical or subcritical) depends upon whether the Froude number is greater than or less than unity. When previously determined values of D_c and V_c are substituted, Eq. (26) indicates that Froude number was in between 0.149 and 0.208 indicating the flow was tranquil (subcritical) and of lower flow regimen as substantiated by the profuse development of small and large scale bedforms in the sandstone bodies. Subcritical flow ($F_r < 1$) is also consistent with field observations of normal fluvial cross beds in paleochannel deposits, whereas antidunes (not observed in study area) form in cohesion-less sedimentary beds under supercritical flow ($F_r > 1$) conditions ([13], p.120). Subcritical flow was considered acceptable for this study based on observed bedforms at the outcrop site and because of it is the most common type of flow in natural streams. The maximum Froude value calculated by putting respective values in Eq. (26) was 0.267, consistent with subcritical flow in all Gondwana sub-basin of Rajmahal basins.

5.6 Channel sinuosity (S_{ch})

The channel sinuosity (S_{ch}) can be indirectly estimated from width and depth of channel and ref. [33] has convincingly developed relationship between them in following regression equation:

$$S_{ch} = 3.5\,(W_c/D_c)^{-0.20} \tag{27}$$

Where W_c is the bankfull width and D_c is the Bankfull channel depth.

When previously estimated values of Wc and D_c are substituted, Eq. (27) indicates that the channel sinuosity was about 1.361 for Brahmini sub basin and progressively increased to 1.632 for the Hurra sub basin through 1.517 for the Pachwara sub basin.

Additional estimates of channel sinuosity can be indirectly from dispersion of paleocurrent directions obtained from the pooled orientation data between channel sinuosity and consistency ratio (L) [41] as

$$\text{Log } S_{ch} = 3.68–0.0684\text{ L} + 0.00032\text{ L}^2 \tag{28}$$

The consistency ratio (L) is concentration measure of orientation vectors and is (magnitude of vector mean/Number of observation) * 100. Paleocurrent analysis of the Barakar sandstone in four sub basins of Rajmahal suggests consistency ratio (L)

82% for the Brahmini, 77% for the Chuperbhita, 74% for the Pachwara and 71% for Hurra respectively. Setting the values of L in Eq. (28) the channel sinuosity comes out to be 1.383, 1.570, 1.663 and 1.728 which signifies low sinuous (braided) for the southern part and moderately sinuous (meandering) in the northern part of the basin. Both Eqs. (27) and (28) yield similar values suggesting that the channel sinuosity increases as the sedimentation progressed from southern to northern part of the Rajmahal Gondwana master basin.

Very recently [37] presents following equation which can be used to estimate the maximum channel sinuosity $P_{(max)}$ of the depositing river:

$$P_{(\max)} = 1.22^{*} \left(900/\sqrt{10}\, D_{50}{}^{0.42}\right)^{0.09} \tag{29}$$

By substituting value for D_{50} (0.40 mm), the maximum channel sinuosity $P_{(max)}$ of the Barakar River in the study area is calculated to be 1.955.

6. Additional parameters

6.1 Paleodischarge (Q_w)

Channel dimensions and paleoslope along with flow velocity permit paleo channel reconstruction using bankfull channel width derived from regression equations and Gibling's scaling factor so as to account variability in channel cross sectional area due to depositional environment as used [6, 42]. A more robust method of estimating paleo-discharge is based on Bagnold's continuity equation as:

$$Q_w = V_c\, A \tag{30}$$

$$Q_w = W_c\, D_c V_c \tag{31}$$

Where Q_w is the discharge and A is the cross-sectional area, which is the product of channel bankfull width (Wc) and channel depth (D_c). V_c is paleoflow velocity was estimated by bedform phase diagram and Darcy Weisbach friction coefficient equation to estimate flow velocity under the assumption that the dominant bedform reflects dominant bed load transport condition during flooding event as described elsewhere. Given bankfull width (W_c), bankfull depth (D_c) and V_c = average flow velocity, Eq. (30) yields a calculated paleo-discharge of 22,070 m^3 for Brahmini sub basin, 12,865 m^3 for Pachwara, 5880 m^3 for Chuperbhita and 4510 m^3 for Hurra sub basin (**Table 1**).

Ref. [43] Related bankfull channel width (W_c) and maximum flow depth (Dc) to bankfull discharge defined as $Q_{2.33}$ based on data collected by ref. [44] for semiarid and sub-humid stable alluvial channels in USA and Australia:

$$Q_{2.33} = 2.66\, W_c{}^{0.9}\, D_c{}^{0.68} \tag{32}$$

On the other hand [45] demonstrate that peak annual flood fluvial discharge is related to drainage basin area (D_A) by the equation:

$$\text{Log}\ Q_{flood} = 2.084 - 0.070\ \log\ D_A{}^2 + 0.865\ \log\ D_A \tag{33}$$

Where Q_{flood} is the discharge with a recurrence interval of 2.33 years and D_A is the drainage basin area. Using previously estimates of (D_A) in this Eq. (33) the mean annual flood is estimated to be about 170,600 m^3/sec for Brahmini sub basin, 112,200 m^3/sec for Pachwara sub basin, 65,460 m^3/sec for Chuperbhita sub basin and 48,640 m^3/sec for Hurra sub basin of Rajmahal Gondwana basin, although there is an order of magnitude uncertainty in this calculation. It is emphasized that this value would not apply year round, would primarily reflect only times of high seasonal yearly discharge. Ref. [45] suggest on the basis of data compiled by them that $Q_{flood}/Q_{average}$ ratios can range over 4 orders of magnitude, from 10 to 10,000. If this is correct than this ratio suggests that average discharge for Barakar Rivers could be as high as 17,060 m^3/sec but as low as 17 m^3/sec. It appears to be that the estimated average discharge of Barakar River compares well with the average discharge of many Indian continental rivers, for example Tapti (6317 m^3/sec), Godavari (3505 m^3/sec), and Indus (6600 m^3/sec).

Ref. [44] used regional hydraulic geometry curves to estimate the discharge of a number of ancient river system based on drainage area. These relationships are developed from analysis of rivers from different climatic and tectonics area. Ref. [4] point out that application to an ancient river system requires some knowledge of the climatic conditions prevailing at that time. The data for this approach require estimates of bankfull channel depth, which can be estimated from empirical relationship as outlined above. Ref. [44] provide hydraulic geometry curves derived using power-law relationships of the form:

$$D_A = a\,A^b \tag{34}$$

$$Q_w = a\,A^b \tag{35}$$

Combining Eqs. (33) and (34) we have

$$D_A = Q_w \tag{36}$$

The Eq. (36) strengthen the views expressed by several workers that in humid and semi-humid regions that area of the drainage basin is approximately equal to the mean annual discharge(Q_w) as recently suggested by workers quoted above. If this is true of the Barakar river system of the study area, the drainage area was about 22,070 km^2 to 4510 km^2. This relationship was used in an attempt to estimate paleoflow parameters of the streams in different parts of world [5, 46, 47].

6.2 Drainage area (D_A)

Drainage basin area is an important component of hydrological analysis which controls sediment supply from provenance to sedimentary basin. Ref. [47] is of the opinion that key information of drainage can be obtained from fluvial deposits, as which are mainly transported from the provenance by fluvial channels. In recent years, a number of studies have investigated river data such as river channel dimensions, bankfull discharge and drainage area generally follow power laws in which bankfull thickness is positively correlated with the drainage area [28, 48]. The power law equations has been used to predict the drainage area of ancient Middle Triassic fluvial deposits of southwest Utah [44], ref. [47] measured bankfull channel-belt thickness of Early Miocene in the Gulf Of Mexico Basin to estimate the drainage areas. Whereas [49, 50] reconstructed drainage basin and sediment routing for the Cretaceous and Paleocene of Gulf of Mexico and Middle Jurassic of northern Qaidam Basin, northwest China respectively by using fluvial scaling relationship. The power law relationship is as:

$$D_A = 249.0\ D_c^{2.44} \tag{37}$$

Where D_c = bankfull channel-belt thickness (bankfull thickness) and D_A = Drainage area. Using this Eq. (37), the drainage area comes out to be 107.4×10^3 km², 86.5 $\times 10^3$ km², 60.52×10^3 km² and 42.2×10^3 km² respectively for the Brahmini, Pachwara Chuperbhita, and Hurra coal basins of Rajmahal Gondwana basin.

Ref. [51] estimate a rivers drainage area based on a power-law relationship between drainage area and peak discharge by the equation:

$$Q_w = k\ D_A{}^m \tag{38}$$

Where Q_w is water discharge in m³/sec, D_A is catchment area (drainage area) in km², *k* constant equal to 0.075 and an *m* exponent of 0.80. Subsequent work by ref. [52] has shown that, the constant and exponent variables of power law relationship (38) change based on different climatic zones as shown in ref. [52] (**Table 1**). As humid climate have been visualized by many workers during deposition of Barakar formation thereby the *k* constant equal to 0.0161 and exponent *m* of 0.9839 have been used. Putting these values in Eq. (38)

$$D_A = 0.0161\ Q_w{}^{0.9839} \tag{39}$$

Where D_A = drainage area and Q_w = mean annual discharge. Substituting the estimated values of Q_w in the above power equation the drainage area was approximately 136.2×10^3 sq. km in Brahmini coal basin and decreased to 28.5×10^3 sq. km to Hurra coal basin as sedimentation progressed in Rajmahal master Gondwana basin.

Barakar sandstone (Permian) drainage networks appear to have been broadly similar in scale to modern intra-basinal rivers draining active mountain belts and the continental-scale drainages, like the Ganga (India-861.4×10^3 km²), the Indus (India-321.2×10^3 km²), Brahmaputra (Bangladesh-194.4×10^3 km²), the Po (Italy-70.0×10^3 km²), and the Rhone (France- 96.2×10^3 km²).

6.3 Stream length (L)

The length of a stream refers to the total length of stream channels in the drainage basin and therefore has units with dimension (L). Theories postulated shows that drainage area and stream length are related in a power function ([53], for references therein). If works done by these workers are correct, a power relation equation can be then derived which would be in form:

$$L = a\ (D_A)^b \tag{40}$$

Where L = stream length, D_A = drainage area and a, b are constants. Ref. [54] derived the values of constants as: a = 1.109 and b = 0.545 while studying relationship between drainage basin area and length of stream for River Gongola, Nigeria. Substituting the constant values in Eq. (40) we have power relation as;

$$L = 1.109\ (D_A)^{0.545} \tag{41}$$

When previously determined value of D_A from Eq. (37) is substituted, Eq. (41) indicates that the length of the Late Paleozoic Barakar River was about 703 km in

Brahmini sub basin and gradually decreasing as channel sinuosity increases through Chuperbhita sub basin (525 km) to Hurra sub basin to about 427 km in Rajmahal Gondwana master basin.

7. Paleo-dimensions, flow dynamics and paleogeography

The quantitative parameters of Early Permian Barakar River in Rajmahal Gondwana basin of eastern India suggest systematic changes in paleo-dimensions and flow dynamics. The Brahmini sub basin in the south was 10.91–14.96 m deep and 817–2956 m wide, carried coarse to medium grained sand (>700 microns) and flowed over a steep northward paleoslope of 6.5 to 16.2×10^{-4} through the alluvial plain; the channel belt width ranging between 6310 and 6780 m and drained an area of 136×103 to 107×10^3 square km. Shields stress, inversion estimates consistently show higher difference in paleoslope (i.e. higher slopes up dip and lower slopes down dip) relative to paleoslope derived from ref. [30] is, perhaps, due to the Trampush's method using a continuous function to estimate slope. Reconstructed Rouse numbers, (Z), indicate that dominant transport modes of bed material varied in space and time (**Table 1**). The proximal field sites consistently exhibit high Z values (between 3.58–3.46), indicating predominant bed load transport. Sediment load parameter value is characteristic of a bed load channel where the bed load forms about formed only about 17.74%, whose mean annual discharge was 22,070 m^3/sec, and mean annual flood was about 170,600 m^3/sec. These quantitative estimates along with sinuosity value of 1.361–1.383 characterize low sinuosity braided stream channels (**Figure 3**). The Flow characteristics remained in the upper part of the lower flow regime (Froude number = 0.208) and flows tend to be dominated by laminar flow because the estimated value of Reynolds number (R_{ep}) is low (**Table 1**); the flow velocity between 1.57 and 1.89 m/sec caused coarse to medium grained sand to be in a dune configuration. The power equations suggest that the bedform length varies from 34.21–25.52 m and corresponding dune height is 1.363 m, comparable in magnitude with large dunes.

As the Barakar River entered in Pachwara basin in the north, the length is reduced to 631 km which drained an area of 81,000 to 86,536 square km; it had a mean annual discharge of 12,865 m^3/sec and a mean annual flood of about 112,200 m^3/sec. The river was about 697 m wide and its depth in between 9.64–13.87 m. This multithread stream with sinuosity of 1.517 carried 15.62 percentage of bed load flowed on a moderate slope of about 0.000144 with flow velocity between 1.35 and 1.77 m/sec. It is evident that the upper part of the lower flow regime caused the medium-coarse grained sand to be in a dune configuration attains a height of 1.247 m and bedform length in between 23.27–31.40 m (large dunes). Calculated high values of Rouse number, Z, indicates that dominant transport modes of bed material and in Pachwara sub basin implying that bed load remains the most important transport mode. A greater width/depth ratio and moderate sinuosity streams are known to transport little silt and clay in their channel perimeter. The trunk Barakar River in this sub basin with modal coarse-medium sand transported its load as bed load at the threshold of motion. Critical stream power in this basin decreases to 0.152 Pa to carry coarser sediments (D_{50} = 0.57 mm) and produce lower amount of sediment flux a common phenomenon in many modern natural rivers. This sub basin shows low paleoslope between 0.000052–0.000152 indicating that river channels carry more wash load relative to total load which, in turn, would imply less boundary shear stress (8.66 Pa) in these river channels. Further north in Chuperbhita sub basin, its length is decreased again to about 525 km, flowed down the paleoslope from southwest to northeast having drainage area of about 60,512–37,000 km^2. The mean annual discharge was about 5880 m^3/sec during normal period and up to 65,450 m^3/sec during periodic flood. Consequently, the water velocity of 1.12 m/sec through most time but increased to 3.01 m/sec during flood and clast flow stages, respectively. The quantitative parameters provide evidence for a relative decline in the channel depth (8.58–12.56 m), width (534 m) and channel belt width (3240–4305 m) and meander lengths of 7230 m, and increase in channel sinuosity up to 1.62–1.66, implying a decline in the slope of the Chuperbhita sub basin as compared with the southerly sub basins (Pachwara and Brahmini). The river channel in this sub basin flowed over a gently sloping surface (0.000139–0.000045) with relatively lower flow velocity of 1.12 m/sec. Indeed, there is an increase in the channel sinuosity owing to an increase in fine clastics and decrease in the stream power as indicated by boundary shear stress and critical shear stress. The sediments so transported are medium to fine sand with more of fine clastics. These changes in the channel parameters together with a decrease in the bed load percentage (10.14) and Rouse number (2.01) are indications of sinuous stream. The hydrodynamic characters (Froude number = 0.174, Reynolds particle number = 29.573) indicated that the flow in the stream channel was tranquil and river flow was in the lower flow regime, which, in turn, gave rise fine- to medium grained sand to be in large and medium scale bed configuration of relatively lower height of 1.073 m and bedform length in between 26.97–18.36 m as compared to southern sub basins.

In the extreme north in Hurra sub basin, its length is further reduced to 427 km and drained an area to the north and northeast of 42,257–28,000 km^2. The average water depth and channel width was 7.25 m, 408 m and increased up to 10.87 and 1380 m, respectively during annual periodic flood. Consequently, the mean annual discharge that was 3890 m^3/sec during normal period may jump up to 48,640 m^3/sec during periodic flood. The water velocity is 0.92 m/sec during normal period and up to 2.84 m/sec during flood season. The stream power is reduced so as the sediment load parameter (M) 6.92 and Rouse Number (Z) further decreased to 1.71, suggestive of the mixed load remains the dominant transport mode of the streams. These narrow streams with meander wavelength of 6576 m swept over a depositional surface sloping at the rate of 116–176 cm/km in northeasterly direction. However, the flow in the stream channel was tranquil and in the upper part of lower flow regimen as indicated by the flow dynamics (Froude number = 0.149, Reynolds particle number = 16.75) which, in turn, facilities development of cross-bedded units in the fine- to medium grained sandstones. These dune bed configurations so developed attains relatively lower height of 0.928 m and bedform length in between 23.33–14.16 m as compared to southern sub basins. The stream channel sinuosity increased up to 1.632–1.728, owing to an increase in the fine clastics, implying a further decline in the paleoslope. However, low sinuous Barakar River, like those of the Brahmini, Pachwara and Chuperbhita became moderately sinuous (? Meandering) streams equivalent to 'mixed load channels', which might be due to intra-basin differential subsidence. To corroborate inferred sediment transport modes, the graphic plots of Reynolds particle number, R_{ep} as a function of Shield stress, τ^* (**Figure 3**), are typical for bed load channel in the Brahmini and Pachwara in the south, whereas the mixed load realm for Chuperbhita and Hurra sub basins in the north. Down-dip, all data straddle the bounds between stability fields for ripples and dunes, which imply unidirectional flow and high to moderate sediment transport rates (both mixed transport and bed load transport).

Early studies have postulated northerly outlet of Permian rivers in this part of Peninsular India based on regional stratigraphic and sedimentary framework and

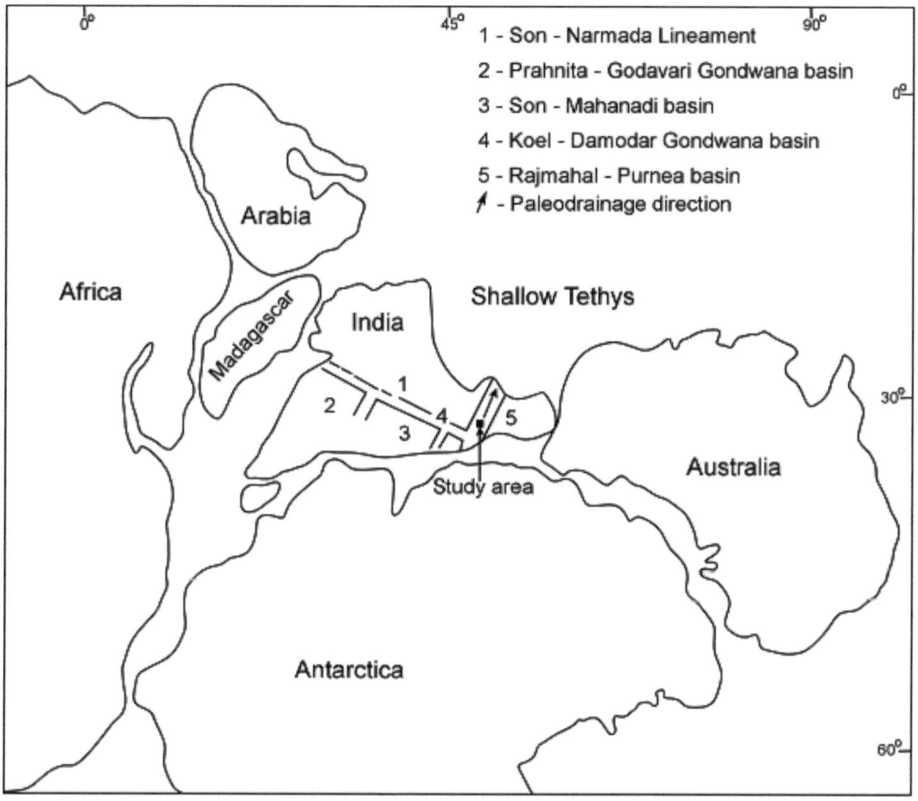

Figure 4.
Paleogeographic reconstruction during early Permian of eastern peninsular India. Solid arrow indicated the outlet of Barakar River.

paleocurrent studies [55, 56]. In addition, newly identified isolated sub basins of Lower and Upper Gondwana litho-units in the sub-surface in Purnea of India and Bogra in western Bangladesh constitute north-south oriented larger Purnea-Rajmahal-Bogra Gondwana Graben [57, 58]. The southerly provenance of subsurface Gondwana rocks in the northern part of Bangladesh, further, justify the northerly sediment transport in this part of eastern India [59]. **Figure 4** illustrates paleogeography of eastern Peninsular India within the framework of eastern Gondwanaland. It is suggested that about 2260 km Long Early Permian River flowed northward from the Fox Ridge and had its outlet towards the cratonic shore zone close to the Himalayan foothills in the north [60].

8. Conclusions

In light of paleo-dimensions and flow dynamics of Early Permian Barakar river of Rajmahal Gondwana sub-basin of eastern India it concluded that:

1. About 2270 km long Early Permian Barakar river of Rajmahal sub-basin of eastern India drained an area of 107,000 to 86,500 sq. km and flowed northward on a steeper slope of 6.5 to 16.2 $\times$ 10^{-4} with a variable flow velocity (157–135 cm/sec) from Fox Ridge; the channel sinuosity was 1.361. The rivers sub-channels

were at least 817 m–700 m wide and 13 m deep. Bed load was moderately high and about 17.74–15.60 of the total load of this river, whose mean annual discharge was 12,800–22,000 m^3/sec during normal period and rose to 112,200–170,000 m^3/sec. These parameters suggest that the stream pattern was multithread (braided) in Brahmini and Pachwara sub-basins in the south. It indicates that the depositing bed load river became mixed- and suspended load within 2260 km in the downstream direction

2. Further northward in Chuperbhita and Hurra sub basins, the river record a decline in paleo-dimensions and flow dynamics such as channel width, channel depth, channel slope, sediment load, discharge and increase in channel sinuosity. Although only 8 to 10 percent of the total river load was bed load; the flow velocity of 92 and 152 cm/sec in the upper part of the lower flow regime caused the fine to medium grained sand to be in dune bed configuration. In the Chuperbhita and Hurra sub basins, it was about 500 km long, drained 42,000 to 62,000 km^2 area to the north and northeast. The gentle slope and abundance of fine clastics suggest slow deposition by single thread river. These paleo dimensions and flow dynamics estimates, indeed, correlate well with the corresponding differences in the lithic- fill of the northern and southern sub basins.

To sum up, the results are merely reasonable approximations that fit the stratigraphic and geologic context in which fluvial facies of the Barakar sandstone deposited in Rajmahal Gondwana sub basin of eastern India.

Symbols and abbreviations

D_{50}	mean particle diameter
ρ_w	fluid water density
H_m	mean dune height
ρs	grain density
S_m	Mean cross bed thickness
τ_b	boundary shear stress
D_c	river bankfull channel depth
τ_{cr}	critical shear stress
D*	maximum channel bankfull depth
g	acceleration due to gravity
W_c	bankfull channel width
f	Darcy-Weisbach friction factor
M	Manning roughness coefficient
L_b	bedform length
Z	Rouse number
M	sediment load parameter
Fr	Froude number
S_a	% silt- clay in channel alluvium
Rep	Reynolds particle number
S_b	% silt-clay in bank alluvium
Q w	water discharge
F	bankfull width/bankfull depth ratio

Q flood	flood discharge
W_{cb}	bankfull width of channel belt
D_A	drainage area
Sc	paleochannel slope
A	area of cross section
R	hydraulic radius
V_c	paleoflow velocity
S_{ch}	channel sinuosity
L	concentration parameter

Author details

Zahid A. Khan[1] and Ram Chandra Tewari[2*]

1 Directorate Geology and Mining U. P. (Retired) 'Khanij Bhawan', Lucknow, India

2 Department of Geology, Sri Jai Narain Post Graduate College (Retired), Lucknow, India

*Address all correspondence to: ram.tewari9@gmail.com

References

[1] Bridge JS, Tye RS. Interpreting the dimensions of ancient fluvial channel bars, channels, and channel belts from wire-line-logs and cores. Bulletin-American Association of Petroleum Geologists. 2000;**84**:1205-1228. DOI: 10.1306/A9673C84-1738-11D7-8645000102C1865D

[2] Leclair SF. Preservation of cross strata due to migration of subaqueous dunes: an experimental investigation. Sedimentology. 2002;**49**:1157-1180. DOI: 10.1046/j.1365-3091.2002.00482.x

[3] Reesink AJH, Van den Berg JH, Parsons DR, Amsler ML, Best JL, Hardy RJ, et al. Extremes in dune preservation: Controls on the completeness of fluvial deposits. Earth-Science Review. 2015;**150**:652-665. DOI: 10.1016/j.earscirev.2015.09.008

[4] Bhattacharya JP, Copeland P, Lawton TF, Holbrook J. Estimation of source area, river paleo-discharge, paleoslope and sediment budgets of linked deep-time depositional systems and implications for hydrocarbon potential. Earth Science Reviews. 2016; **153**:77-110. DOI: 10.1016/j.earscirev.2015.10.013

[5] Lyster SJ et al. Reconstruction the morphologies and hydrodynamics of ancient rivers from source to sink: Cretaceous western interior basin, Utah, USA. Sedimentology. 2021;**68**: 2854-2886. DOI: 10.1111/sed.12877

[6] eBhattacharya JP, Tye RS. Searching for modern Ferron analogs and application to sub-surface interpretation. In: Adams TCCRD, Morris TH, editors. Analog for Fluvial-Deltaic Reservoir Modeling: Ferron Sandstone of Utah. Vol. 50. Amer. Assoc. Petroleum Geologists, Studies in Geology; 2016. pp. 39-57

[7] Ganti V, Whittaker A, Lamb MP, Fischer WW. Low gradient, single threaded rivers prior to greening of the continents. Proceedings of National Academy Science. 2019;**116**:11652-11657. DOI: 10.1073/pnas.1901642116

[8] Khan ZA, Tewari RC. Paleocurrent, paleohydraulic and paleo-geography of Miocene-Pliocene Siwalik foreland basin of India. In: Advances in Geology. Vol. 2015. Hindawi Publishing Corporation; 2015. DOI: 10.1155/2015/968573. Article ID 968573

[9] Veevers JJ, Tewari RC. Gondwana master basin of Peninsular India between Tethys and the Interior of the Gondwanaland province of Pangaea. Vol. 187. Geological Society America; 1995

[10] Khan ZA, Tewari RC. Quantitative model of Early Permian coal bearing Barakar cycles from Son-Mahanadi and Koel-Damodar Gondwana coalfields of eastern-central India. Vol. 9. Gondwana Geological Magazine; 2007. pp. 115-125

[11] Khan ZA, Tewari RC, Hota RN. Facies analysis, Markov model and linking of sub environments in the early Permian Barakar coal measures of Godavari Gondwana Basin of southeastern India. 2020. Journal of Geological Society India. 2020;**95**:599-608. DOI: 10.1007/S12594-020-1486-0

[12] Mohrig D, Heller PL, Paola C, Lyons WJ. Interpreting avulsion process from ancient alluvial sequences: Guadalope- Matarranya system (northern Spain) and Wasatch Formation (western Canada). Bulletin-American Association of Petroleum Geologists. 2000;**112**:1787-1803

[13] Bridge JS. Rivers and Floodplains: Forms, Process, and Sedimentary

Record. Hoboken: John Wiley & Sons; 2009. p. 542

[14] Leclair SF, Bridge JS. Quantitative interpretation of sedimentary structures formed by river dunes. Journal of Sediment Research. 2001;**71**:713-716. DOI: 1527-1404/01/071-713/S 03.00

[15] Adams MM, Bhattacharya JP. No change in fluvial style across a sequence boundary, Cretaceous Blackhawk and Castigate Formation of central Utah, USA. Journal of Sediment Research. 2005;**75**:1038-1051. DOI: 10.2110/jsr.2005.080

[16] Lin W, Bhattacharya JP, Campbell C. Temporal evolution of fluvial style within a compound incised valley, The Ferron Notom Delta, Henry Mountain Region, Utah. Journal of Sediment Research. 2010;**80**:529-549

[17] Allen JRL. A quantitative model of grain size and sedimentary structures in lateral deposit. Journal Geology. 1970;**7**: 129-146

[18] Wu C, Bhattacharya JP, Ullah MS. Paleohydrological 3-D facies architecture of ancient point bars, Ferron Sandstone, Norton delta, south-central Utah, USA. Journal of Sediment Research. 2015;**85**: 399-418. DOI: 10.2110/jsr.2015.29

[19] Ethridge FG, Schumm SA. Reconstructing paleochannel morphologic and flow characteristics: methodology, limitations, and assessment. In: Miall AD, editor. Fluvial Sedimentology. Vol. 5. Canadian Society of Petroleum Geologists; 1977. pp. 703-722

[20] Bradley RW, Venditti JG. Re-evaluating dune scaling relations. Earth-Science Reviews. 2017;**165**:356-376. DOI: doi.org/10.1016/j.earscirev.2016.11.004

[21] Mahon RC, McElroy B. Indirect estimation of bed load flux from modern sand-bed rivers and ancient fluvial strata. Geological Society of America. 2018;**15**:738-748. DOI: 10. 1130/G40161.1

[22] Ashley GM. Classification of large-scale subaqueous bedforms; A new look at an old problem. Journal of Sedimentary Research. 1990;**60**:160-172

[23] Gualtieri C, Martone IF, (Jr), N. P. and Ianniruberto, M. Bedform morphology in the area of the confluence of the Negro and Solimoes-Amazon Rivers, Brazil. Water. 2020;**12**:1-23. DOI: 10.3390/w12061630

[24] Castelltort S. Empirical relationship between river slope and the elongation of bars in braided rivers: A potential tool for paleoslope analysis from subsurface data. Marine and Petroleum Geology. 2020;**112**:1-13

[25] Bridge JS, Mackey SD. A revised alluvial stratigraphy model. In: Marzo M, Puigdefabregas C, editors. Alluvial Sedimentation. 1993. pp. 317-336. DOI: 10.1002/9781444303995.ch22

[26] Fielding CR, Crane RC. An application of statistical modeling to the prediction of hydrocarbon recovery factors in fluvial reservoirs sequences. In: Ethridge FG, Flores RM, Harvey MD, editors. Recent developments in Fluvial Sedimentology. Vol. 39. SEPM Special Publication; 1987. pp. 321-327

[27] Schumm SA. River adjustment to altered hydrologic regimen—Murrumbidgee River and paleochannel, Australia. In: U. S. Geological Survey Prof. Paper. Vol. 598. 1968. p. 65. DOI: 10.3133/pp598

[28] Blum MD, Martin J, Milliken K, Garvin M. Paleo-Valley systems: insights from Quaternary analogs and

experiments. Earth-Science Reviews. 2013;**116**:128-169. DOI: 10.1016/j.earscirev.2012.09.003

[29] Shield A. Application of similarity principles and turbulence research to bed load movement. [Anwendungt der Achnlichkeitsmechanik und der Turbulenzforschung auf die Geschiebebewegung, Mitteilung der Preussischen Versuchsanstalt furWasserbau undSchiffbau, Heft 26, Berlin]. 1936. p. 26

[30] Trampush SM, Huzurbazar S, McElroy B. Empirical assessment of theory for bankfull characteristics of alluvial channels. Water Resources Research. 2014;**50**:9211-9220. DOI: 10.1002/2014WRO15597

[31] Andrews ED. Bed-material entertainment and hydraulic geometry of grave-bed Rivers in Colorado. Geological Society of America Bulletin. 1984;**95**:371-378. DOI: 10.1130/0016-76061984)95:BEAHGO>2.0.CO

[32] Ninke DJ, Evans JE. Alluvial architecture of the Early Pennsylvanian Sharon Formation in northeastern Ohio. Ohio Journal of Science. 2002;**102**:70-81

[33] Lynds RM, Mohrig D, Hajek EA, Heller PL. Paleoslope reconstruction in sandy suspended load dominant rivers. Journal of Sediment. Research. 2014;**84**:825-836. DOI: 10.2110/jsr.2014.60

[34] Kleinhans MG. Flow discharge and sediment transport models for estimating minimum timescale of hydrologic activity and channel and delta formation on Mars. Journal of Geophysical Research. 2005;**110**:E 12003. DOI: 10.1029/2005 JE002521

[35] Wohl E. Mountain rivers. Water resources monograph. American Geophysics Union, Washington. 2000; **14**:70-85

[36] Rubin DM, McCulloch DS. Single and superimposed bedforms—A synthesis of San Francisco Bay and Flume observations. Sedimentary Geology. 1980;**26**:207-231. DOI: 10.1016/0037-0738(80)90012-3

[37] Chen L, Ji H, Zhan L. Sinuosity of alluvial meandering rivers with scroll bars. Arabian Journal of Geosciences. 2021;**15**:111-138. DOI: doi.org/10.1007/S 12517-021-09318-y

[38] Ferguson RL, Church M. A simple universal equation for grain settling velocity. Journal Sedimentary Research. 2004;**74**:933-937. DOI: 10.1306/051204740933

[39] Dade WB, Friend PE. Grain size, sediment transport regime, and channel slope in alluvial rivers. Journal of Geology. 1998;**106**:661-676. DOI: 10.1086/516052

[40] Allen JRL. Sedimentary Structures: their character and physical basis. In: Development of Sedimentology, 30B. Vol. II. Amsterdam: Elsevier Publ. Company; 1982. p. 663

[41] Ghosh P. Estimation of channel sinuosity from paleocurrent data: A method using fractal geometry. Journal of Sediment Research. 2000;**70**:449-455

[42] Abatan OO, Weislogel A. Paleohydrology and machine-assisted estimation of paleo-geomorphology of fluvial channels of the lower middle Pennsylvanian Allegheny formation, Birch River, West Virginia. Frontiers in Earth Science. 2020;**7**:1-20. DOI: 10.3389/feart.2019.00361

[43] Williams GP, Rossman PI, Zimbelman JR. Evaluation of

paleohydrologic models for terrestrial inverted channels: Implications for application to Martian sinuous ridges. Geomorphology. 2009;**107**:300-315

[44] Schumm SA. Fluvial paleochannel. In: Rigby JK, Hamblin WK, editors. Recognition of Ancient Sedimentary Environments. Vol. 16. Society of Economic Paleontologists and Mineralogists, Special Publication; 1972. pp. 179-183

[45] Mulder T, Syvitsky JPM. Turbidity currents generated at river mouths during exceptional discharges to the world's oceans. Journal Geology. 1995; **103**:285-299

[46] Davidson SK, North CP. Geomorphological regional curves for prediction of drainage area and screening modern analogue for rivers in the rock record. Journal Sediment Research. 2009;**79**:773-792. DOI: 10.2110/jsr.2009.080

[47] Blum MD, Womack JH. Climate change, sea-level change and fluvial sediment supply to deepwater systems. In: Keller B, Martinsen OJ, McCaffery B, editors. External Controls on Deep Water Depositional Systems: Climate, Sea-Level and Sediment Flux. Vol. 92. SEPM Special Publication; 2009. pp. 15-39

[48] Anderson JB, Wallace DJ, Simms AR, Rodriguez AB, Weight RW, Taha ZP. Recycling sediments between source and sink during a eustatic cycle: Systems of Late Quaternary northwestern Gulf of Mexico basin. Earth Science Review. 2016;**153**:111-138. DOI: 10.1016/j.earscirev.2015.10.014

[49] Milliken KT, Blum MD, Snedden JW, Galloway WE. Application of fluvial scaling relationships to reconstruct drainage basin evolution and sediment routing for the cretaceous and Paleocene of the Gulf of Mexico. Geosphere. 2018;**14**:749-767

[50] Liu B, Shao Q, Wang LY, Li XT, Y. N. and Xu, J. Application of channel-belt scaling relationship to middle Jurassic source to sink system in the Saishiteng area of the northern Qaidam basin, Northwest China. Journal Paleogeography. 2019;**8**:581-586. DOI: 10.1186/s 42501-019-0031-9

[51] Syvitsky JPM, Milliman JD. Geology, geography and humans battle for dominance over the delivery of fluvial; sediment to the coastal ocean. Journal of Geology. 2007;**115**:1-19

[52] Eide CH, Muller R, Helland-Hansen W. Using climate to relate water discharge and area in modern and ancient catchments. Sedimentology. 2018;**65**:1378-1389. DOI: 10.1111/sed.12426

[53] Carlton RO. Fundamentals of fluvial geomorphology. 1st ed. Rout-ledge Publ; 2007. p. 280

[54] Oyegoke SO, Ifeadi C. Relationship between drainage area and stream length for river gondola, Nigeria. Pacific Journal Science and Technology. 2007;**8**:184-189

[55] Tewari RC, Casshyap SM. Paleoflow analysis of Late Paleozoic Gondwana deposits of Giridih and adjoining basins, Bihar and paleogeographic implications. Journal Geological Society of India. 1982; **23**:67-79

[56] Tewari RC, Veevers JJ. Gondwana basins of India occupy the middle of a 7500 km of radial valleys and lobes in eastern-central Gondwanaland. In: VIII International Gondwana Symposium, Australia: A.A.Balkema, Rotterdam; 1993. pp. 507-512

[57] Alam M, Curray JR, Rahman Chowdhury ML, Gani MR. An overview of the sedimentary geology of Bengal Basin in relation to the regional tectonic framework and basin-fill history. Sedimentary Geology. 2003;**155**:179-208

[58] Islam MR, Hayashi D. Geology and coal bed methane resource potential of the Gondwana Barapukuria coal basin, Bangladesh. International Journal of Coal Geology. 2008;**75**:127-143. DOI: 10.1016/j.coal.2008.05.008

[59] Alam MI et al. Late Paleozoic detrital history of eastern Gondwanaland: Petrofacies and detrital geochronology of Permo-Carboniferous intra-cratonic sequences of the northwest Bengal basin. Journal of Sedimentary Research. 2020;**90**:389-402

[60] Holbrook J, Wanas H. A fulcrum approach to assessing source-to-sink mass balance using channel paleohydrologic parameters derivable from common fluvial data sets with an example from the cretaceous of Egypt. Journal of Sediment. Research. 2014;**84**:349-372. DOI: 10.2110/jsr.2014.29

Chapter 4

Sedimentary Rock Fabric Characterization

Ishaq Yusuf and Eswaran Padmanabhan

Abstract

The goal of rock fabric characterization is to describe the spatial and geometric distribution of pore attributes as they impact petrophysical parameter variations such as porosity, permeability, and water saturation. Pore attributes such as pore size and pore volumes are critical petrophysical properties as they change and vary over a short distance or side-by-side, and determine the exploitable capacity of sedimentary reservoir rocks. A proposed new insight is in the systematic characterization of fabrics by the combination of both digital petrographic scanning and conservative petrographic description, as we show that multiple fabrics occur in a single lithofacie in the form of a fabric domain. Characterization shows that these domain types determine the spatial and geometric distribution of the variable pore sizes and volume (porosity) within a certain lithofacie fabric. Based on this, we also infer that these fabric domain types are responsible for the multiple occurrences of hydraulic fluid units (HFU's) and anomalous porosity-permeability relationships in siliciclastic sandstone sedimentary rocks since no long-range process aligns the microscopic internal fabric or microfabric architecture among grain aggregates in sedimentary rock.

Keywords: sedimentary rock, rock fabric, fabric domain, reservoir characterization, pore size, petrographic description

1. Introduction

The goal of rock fabric characterization is to describe the spatial and geometric distribution of pore attributes as they impact petrophysical parameters such as porosity, permeability, and water saturation. We showed that multiple fabric domains exist in lithofacie at microscale instead of the traditional occurrence of a single fabric in a lithofacie. In this chapter, we expand and showed new insight into rock fabric characterization of *fabric domains* to pore size and porosity distribution, that is, the spatial distribution of pore size and porosity within the different fabric domain types, and show how their distribution can be linked to fabric domain types.

Petrographic investigation is one of the key techniques in sedimentary lithofacie fabric description/characterization, before advanced application of higher-resolution microscopy such as the scanning electron microscope (SEM) or Fourier transform scanning electron microscopy (FESEM) and transmission electron microscope (TEM) to visualize sedimentary rock fabric in their undisturbed state at the nanoscale (micrometer or nanometer) especially for argillaceous sediments [1]. It is essential to

IntechOpen

improve on understanding of the rock fabric that controls overall reservoir behavior and to advance by deviating from the orthodox petrographic reservoir description. Petrographic characterization is marginally described as the different grain patterns (*viz*. grain assemblage (packing), grain shapes, minerals, rock fragments, grain orientation, grain sorting) are not completely visualized as a whole under the polarized microscope. Variations in the sandstone grain assemblage vary in sedimentary rock, and so in their petrophysical and engineering properties that potentially differ among overall response to stress, deformation, diagenesis, compaction, fracturing, and fluid flow at microscale level. This chapter presents scanning of petrographic slides (thin section) to have a wholly view or panoramic visualization of grain assemblage and pattern toward identification of different domain of fabric types before the conventional detail petrographic description or characterization. This act will extend knowledge of understanding macroscopic reservoir behavior as to what is responsible at the microscopic scale for petroleum geoscientist and engineers. Also, some technical question will be understood, for example, why do multiple hydraulic fluid units (HFU's) exist in a single sandstone lithofacie? And question such as why a characterized porous geologic material with a high porosity exhibits poor or low fluid flow potential (permeability) even when investigated with a high-pressure gas equipment, for example, helium porosimetry? This chapter relates rock fabric domain to pore sizes and pore volume (porosity) distribution as a most critical to hydrocarbon reservoir development and production efficiency.

1.1 Schemes for fabric classification

Schemes for fabric classification began with carbonate classification by Archie [2] on carbonate reservoir rock in relation to pore size and fluid distribution (petrophysical properties). Choquette and Pray [3] present a classification of porosity which stresses interrelations between porosity and other geologic features in sedimentary carbonates. Rock-fabric and carbonate pore-space classification for reservoir characterization were developed by Lucia [4]. In argillaceous sediments, a quantitative scanning electron microscopic (SEM) methods are used for soil fabric analysis [5] and Sokolov and O'Brien [6] carried out fabric classification system for argillaceous rocks, sediment, and soils based upon the microfabric shown by the scanning electron microscopy (SEM) as it provides a frame of references in describing microfabrics. The nomenclature of pedological features and the various levels of structure and fabric are developed by Brewer and Sleeman [7]. In hard rocks, fabrics is used to characterize sediment transport, magma flow, and dynamo-metamorphic deformation using polar plots [8].

Not until the thin-section technique was accessible, there are no serious study of the fabric and composition of sandstones and utilization of their microscopical characteristics to elucidate the natural history of the rock [9]. Sedimentary rock fabric has been defined, studied, and documented by several renowned sedimentologists [10–12] as a mutual arrangement of grains in sediment including orientation of grains and their packing. Relationships between grains and matrix are examined and characterized as either of clasts-supported or matrix-supported. Earlier sedimentologists [10–12] are of the opinion that the fabric of a lithofacies can be characterized as aforementioned, while emphasizing on other rock fabric physical features such as grain orientation, packing, and sorting to better understand the description of the lithofacie fabric under investigation. The new insights, while upholding the present fabric classification by early sedimentologists, is to view entirely a lithofacie fabric, and to examine and visualize potential multiple fabrics instead of a single fabric classification and description as earlier sedimentologist outlined in many books [10–12]. This approach in this chapter presents a systematic fabric classification for siliciclastic sandstone thereby maintaining existing classifications by earlier sedimentologists.

1.2 Concept of fabric domain characterization

The fabric of a rock includes the complete spatial and geometric configuration of all its components and encompasses (micro-) structure and crystallographic preferred orientation of the ensemble of minerals [13]. Microfabric is referred to as arrangement of soil aggregates [14]. The importance of microfabric in the development of sedimentary deposit was suggested by Sorby (1908). Terzaghi (1925) and Casagrande (1932) proposed primitive fabric model to help explain the relationship and sensitivity of cohesive sediment. In Argillaceous sediments (shale), fabric produced by modern sedimentary environment has been recorded by using scanning electron microscope (SEM), transmission electron microscope (TEM), and thin section technique [15]. Electron microscopy observations [16–19] have revealed the presence of multiple particles (domains) as the predominant fundamental particle type rather than the thin, single-plate particles proposed in early models [1]. A domain is considered to have significant structural integrity and to behave in a functional sense as a unit particle for a finite period of time under an applied stress regime. Thus, it is important to note that "a particle" can be defined in terms of its morphology as well as its function [15]. By an online dictionary, a *fabric domain* is the three-dimensional area of the rock occupied by uniform fabric and is delineated by some kind of boundary such as structural or compositional discontinuity (http://encyclopedia2.thefreedictionary.com/fabric+domain).

In sandstones, microfabrics are mixtures of close-packed domains and packing flaws [20]. Close-packed domains contain relatively smaller pores and pore throats, and so have affinity for the wetting phase, while the losses-packed domain have been demonstrated to possess larger pores connected by larger pore throat, for they form pathways for fluid flow and virtually all the non-wetting phases [20]. The close-packed zones have smaller pores and pore throats, and along with microporosity, which retains irreducible water. The aforesaid description of a fabric domain is in sandstone is further extended to share more insights to the present day understanding and challenge to provide technical explanation for reservoir rocks efficiency or under performances that are constantly linked to rock fabrics. It is undisputed that the fabrics of reservoir rock determine storage capacity and fluid displacement within the framework. As rock physical properties *viz.* grain size, shape, sorting, and orientations to mention but few are the essential elements that shape the behaviors and effectiveness of any sandstone reservoir rock.

A fabric domain in this context is considered as an aggregate unit that exhibits a functional physical textural configuration (variable grain shape, orientation, sorting, contact, matrix composition, and compaction) and pore attribute properties (pore volume, pore size, pore throat, pore structure, and pore interconnectivity) that controls fluid flow dynamic and other petrophysical properties (e.g., water saturation, wettability, thermal conductivity). This description of a domain in this context, however, upholds earlier given domain characteristic for both the close-packed and packing flaws [21]. Considering thin-section micrographs for conservative description and characterization will not give a full or wholly view of the entire rock fabric (**Figure 1**), such that some vital information will be lost to understand the macroscopic behavior

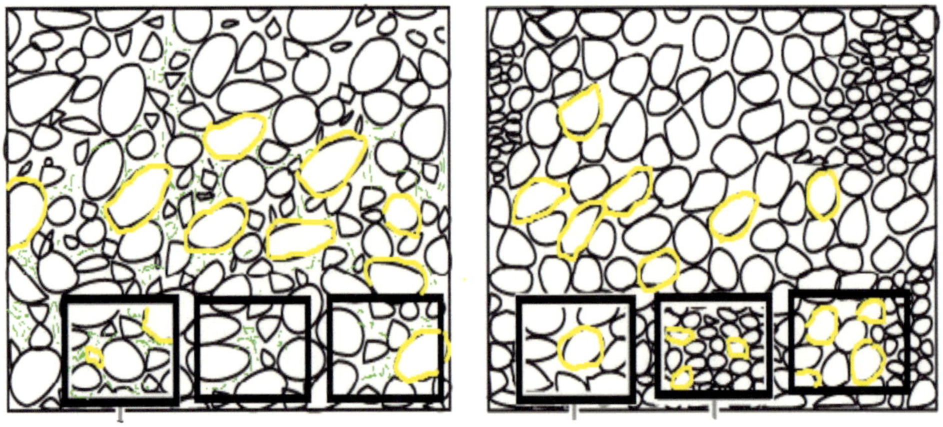

Figure 1.
Schematic diagram of scan thin section micrographs depicting both homogeneous and heterogeneous fabric domains, while the black square boxes depicted polarized microscope pin-hole captured image for orthodox description and characterization into clasts-supported fabrics.

or efficiency of the rock. The above schematic diagram (**Figure 1**) reveals two fabric domain types, instead of the conservative characterization as a clasts-supported fabric types on which a detail characterization of fabric physical properties such as grain sorting, packing, contacts can be carried upon for better description.

1.2.1 Fabric domain mapping

The proposed mapping of fabric domains includes scanning of prepared thin-section micrograph at a resolution scale of about 5 mm to have clearer image resolution (**Figures 2–6**). This will enable to visualize completely all the different grain assemblages or colonies at a single glance. It is also achievable when a thin-section micrograph is systematically gridded to capture images under a polarized microscope, but will be cumbersome when attempting to join all captured micrograph images

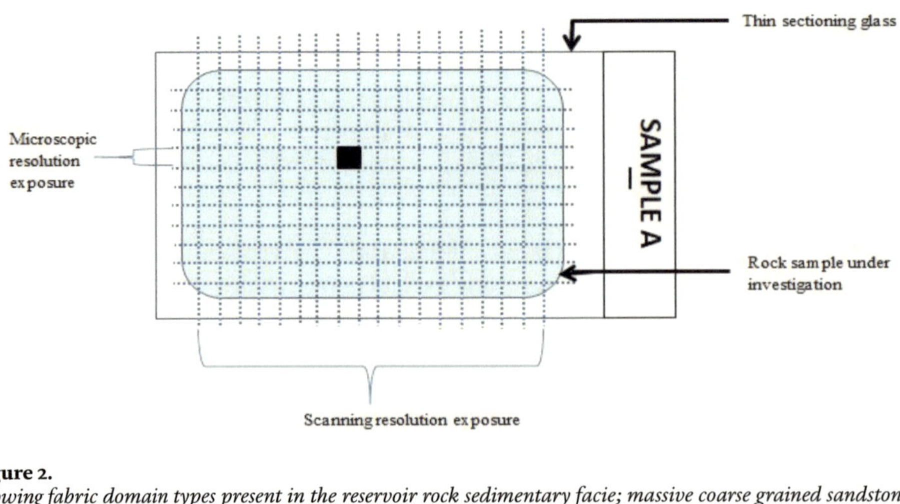

Figure 2.
Showing fabric domain types present in the reservoir rock sedimentary facie; massive coarse grained sandstone (MCGS) from Offshore, Malaysia. In yellow dotted-line is the Random domain, in red dotted-lines in the Clasts-supported domain and in blue dotted-line is the Matrix- supported domain.

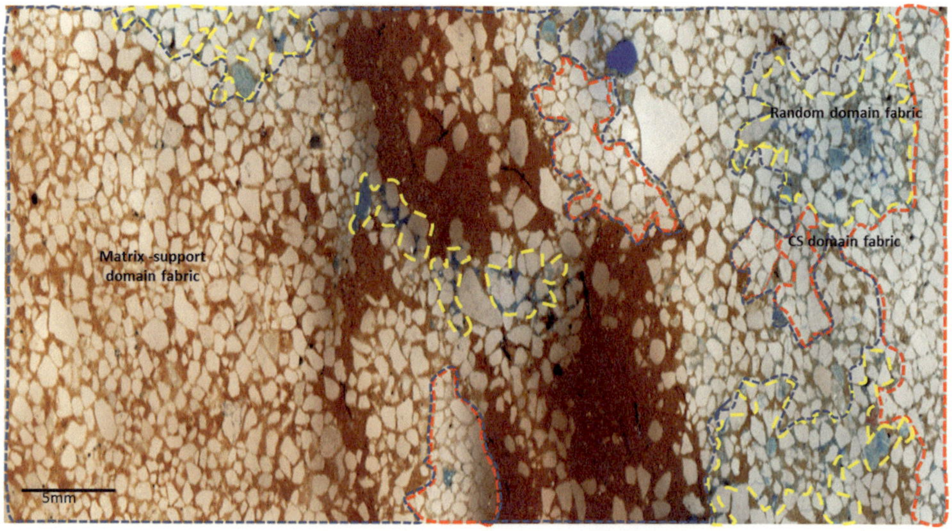

Figure 3.
Showing fabric domain types present in the reservoir rock sedimentary facie; massive Burrowed medium grained sandstone (BMGS) from the Offshore, Malaysia. In yellow dotted-line is the Laminar/matrix supported domain, in red dotted-lines in the Clasts-supported domain, in black dotted-line is the fractured-dominated domain and in purple dotted-line is the random domain.

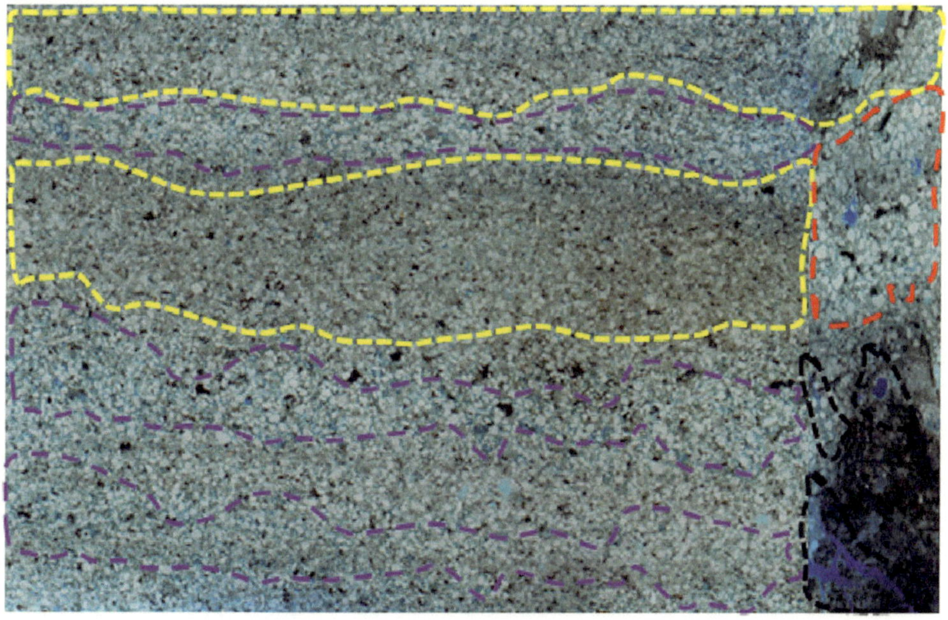

Figure 4.
Showing fabric domain types present in the reservoir rock sedimentary facie; parallel laminated friable fine grained sandstone (PLFFGS) from Offshore, Malaysia. In yellow dotted-line is the Matrix supported domain, in red dotted-lines in the Clasts-supported/Fractured-dominated domain, and in purple dotted-line is the random domain.

at 500-μm magnification into a single micrograph due to edges-effect difference as can be easily be achieved using a high-resolution scanning machine (**Figure 7**). The scan micrograph image carefully visualizes all grain assemblages or pattern nature

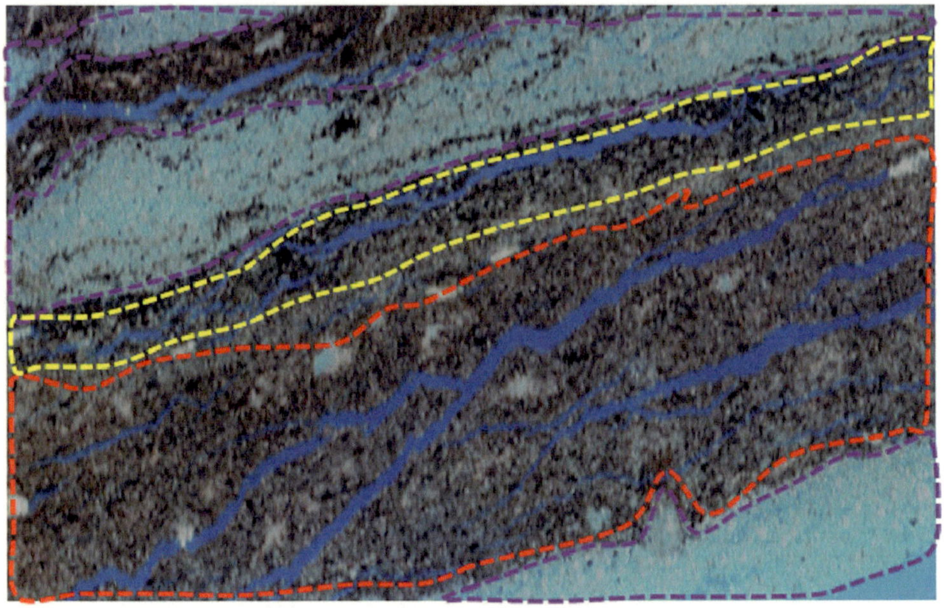

Figure 5.
Showing fabric domain types present in the reservoir rock sedimentary facie faintly laminated fine grained sandstone (FLFGS) from, Offshore, Malaysia. In yellow dotted-line is the Random/fractured-dominated domain, in red dotted-lines in the matrix-supported domain and in deep blue dotted-line is the Clasts-supported domain.

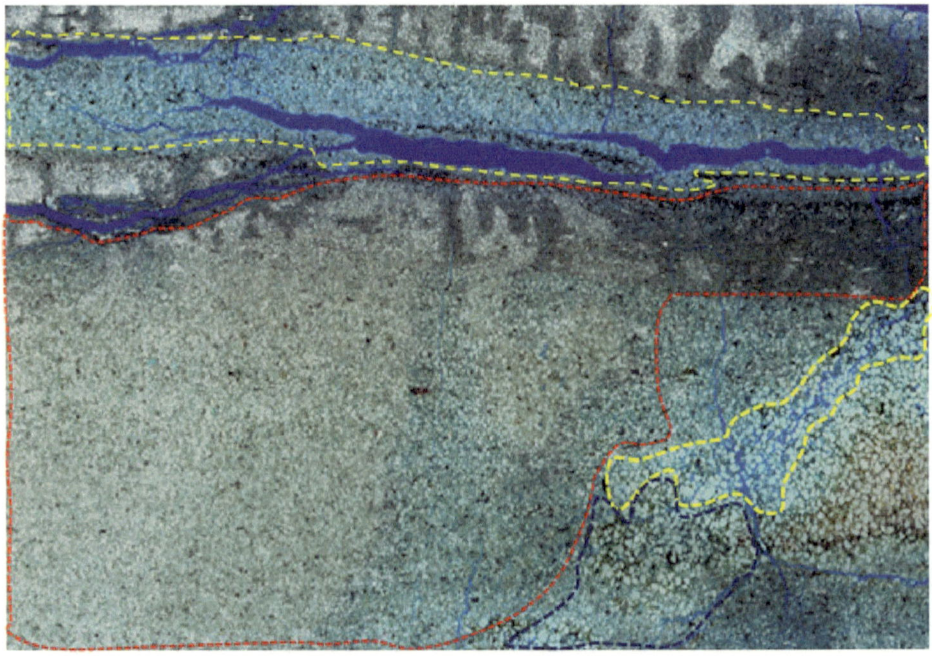

Figure 6.
Showing fabric domain types present in the reservoir rock sedimentary facie massive very fine-grained sandstone (MVFGS) from Offshore, Malaysia. It exhibited only two (2) fabric type Matrix-supported and Fractured-dominated domains.

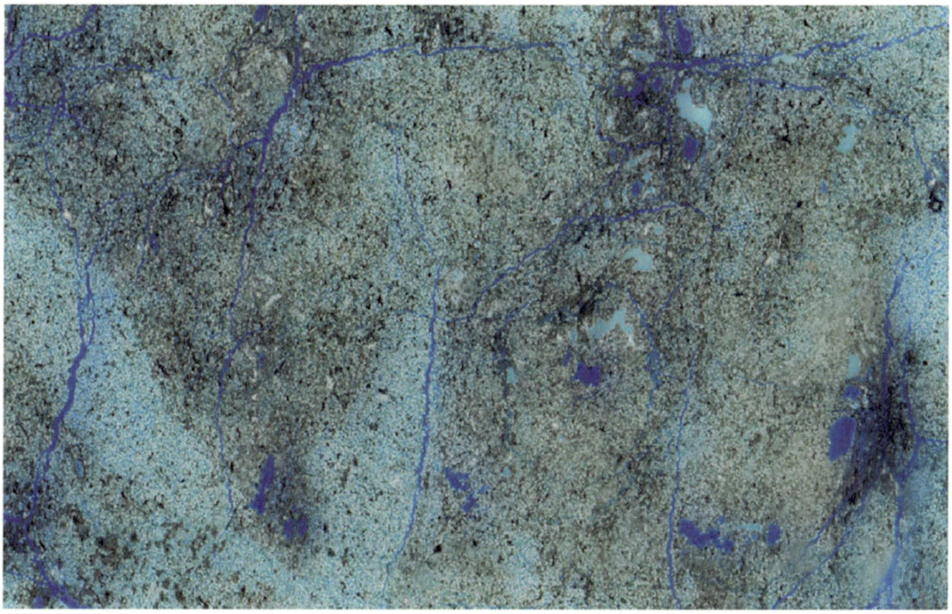

Figure 7.
Schematic diagram showing study area under the microscope and scan area for fabric domain type mapping.

with reference to the dominant grain packing, contacts, sorting, orientation, matrix distribution, compaction and intergranular pore-spaces will be characterized into fabric-type domain (**Figures 2–6**).

In a practical approach to achieve mapping of the variable domains in a sedimentary facie, each representative micrograph (s) is scanned using a high-resolution scanning machine as earlier mentioned. The output gives a panoramic view or coverage of the entire aggregates of geometric and spatial configuration on the micrograph; thereby, a careful observation of the different domain fabric type will be mapped out clearly (**Figures 2–6**).

1.2.2 Fabric domain types

The reservoir sandstone facies of the West Baram Delta as an example is characterized into five fabric domains as in subsection 1.2.1 based on the dominant textural features (grain size, shape, grain sorting, grain-contact, grain orientation, pore space, and matrix content) namely the clasts-support, random, matrix-support, and laminar and fractured-dominated fabrics. The clasts-support (types 1 & 2), matrix-support (types 1 & 2), and random and fractured-dominated fabric (types 1 & 2). Subtype (type 1 & 2) for the clasts-support, matrix support, random and fractured-dominated are grouped based on their variances in percentage composition of matrix content (**Table 1**).

From the above fabric domain-type mapping, it indicates that a single thin micrograph is not ideal to fully represent a single rock fabric type. This being that there are variable and discrete variations in physical grain assemblages and perhaps their properties within the micrograph. Thus, changes and variations will occur within the developed different fabric domains across the entire micrographs. It occurs that the multiple fabric domain type facilitates existence of multiple hydraulic fluid units (HFU's) [22] orchestrated mainly by the variable difference or in some case similar

Lithofacies	Fabric types	Fabric type characteristics						
		Grain size (%)	Grain shape	Grain sorting	Grain packing	Matrix (%)	Grain composition	Porosity
MCGS (15)	Clasts-supported (Type 1)	48.87	Angular-sub rounded	Moderate	Point, moderate long/concavo convex	Minor@7.78	Moderate	High
	Random	48.18–51-52	Angular-sub rounded	Poorly-moderate	Point, moderate long/concavo convex	Minor@6.10	Moderate	High
	Matrix-supported	48.32	sub angular-sub rounded	Moderate	Point, moderate long/concavo convex	Moderate @37.70	Very low	Very low
BMGS (24)	Laminar	Medium @ 32.30–53.20	sub angular-sub rounded	Moderate	Long & concavo-convex	High @42.30	Moderate	Moderate
	Clasts-supported (Type 2)	Medium @ 53.10	Moderate angular-sub rounded	Moderate - poorly	Float & point	Moderate @38.10	Moderate	Moderate
	Fractured-dominated (Type 1)	Medium @ 48.02	Moderate angular to Sub rounded	Well- sorted	Long, float & point	Minor @9.10	Minor	High
	Random	Medium @ 54.22	Sub rounded	Moderate – well- sorted	Long, float & point	Minor @5.60	Moderate	Moderate high
FLFGS (36)	Fractured-dominated (Type 2)	Fine @ 39.80	Sub rounded	Poorly	Long, float & point	Moderate @ 38.40	moderate	Moderate High
	Random	Fine @ 52.30	Sub rounded-rounded	Moderate - poorly	Long, float & point	Minor @56.60	low	high
	Clasts-supported (Type 2)	Fine – medium @ 37.60	Moderate angular to sub rounded	Poorly	Float, long – concavo convex	High @42.10	Moderate	High
	Matrix-supported	Fine medium @48.30	sub angular-sub rounded >100°	poorly	Float, long – concavo convex	High @35.10	Moderate	Very High
	Laminar	Fine coarse @53.20	Sub angular - sub rounded	Well-sorted	Point, long-concavo-convex	Moderate@23.10	Moderate	low

DOI: http://dx.doi.org/10.5772/intechopen.108046

Lithofacies	Fabric types	Fabric type characteristics						
		Grain size (%)	Grain shape	Grain sorting	Grain packing	Matrix (%)	Grain composition	Porosity
PLFFGS (37)	Clasts-supported	Fine @33.20	Angular-sub angular	Very poorly	Long-concavo convex	Moderate@43.30	Moderate	Low
	Matrix-supported	Fine @ 26.12	Angular-sub angular	Very poorly	Float-concavo convex	High 47.80	Moderate	Very low
	Fractured-dominated	Fine @43.30	Sub rounded - rounded	Moderate	Float-point	High	Moderate	Moderate High
	Random	Fine @45.30	Angular-sub angular	Moderate	Point, long-concavo-convex	High	Moderate	Very low
MVFGS (20)	Matrix-supported	Fine @42.20	Angular-sub angular	Moderate	Point, long-concavo-convex	High	Moderate	Moderate
	Fractured-dominated	Fine medium @53.28	Angular - sub angular	Moderate	Float-point	Moderate	Moderate	High

Table 1.
Fabric types compositional characteristics in sandstone from Offshore, Malaysia.

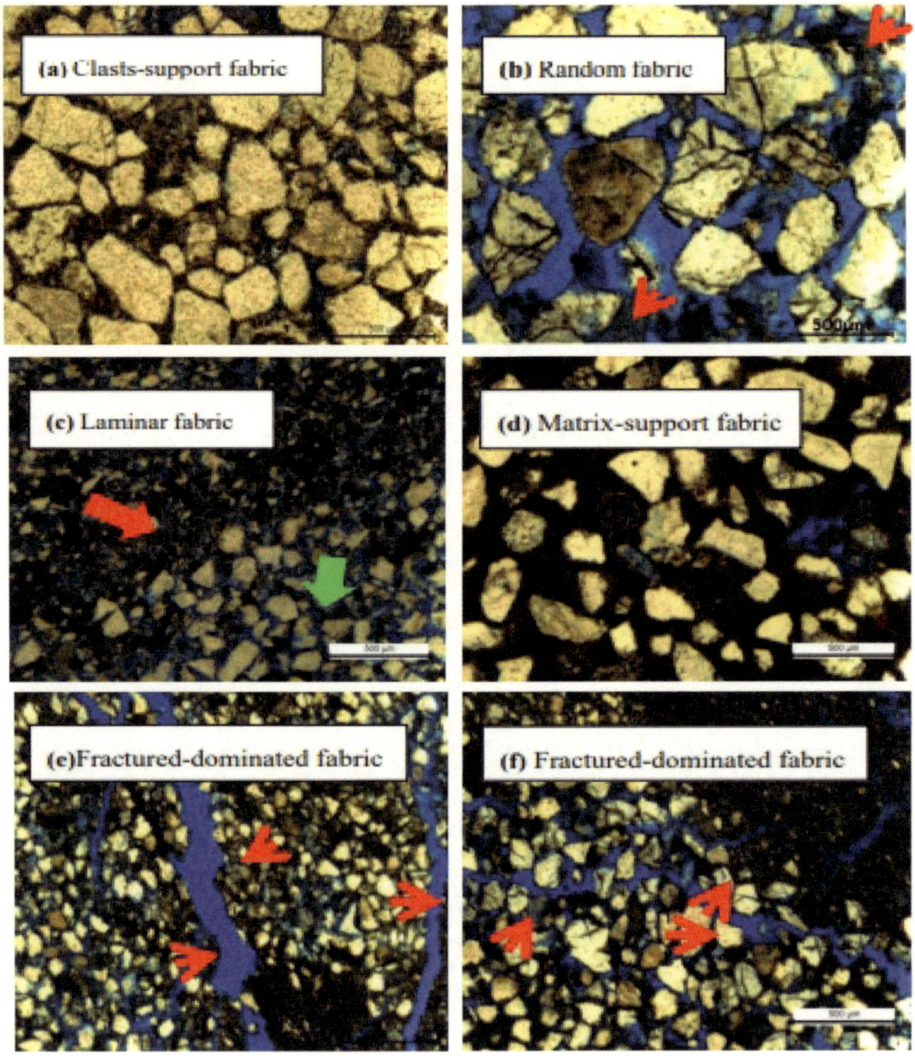

Figure 8.
Showing the mapped different fabric domain types at 100 μm and 500 μm magnification (a) Clasts-supported grains @ depth 987.3 m; (b) Random domain grains are pushed apart with minor matrix, sample @ depth 1943 m (c) Laminar @ 972.6 m (d) Matrix-support, matrix filling intergranular pore space; sample collected @ depth 1509.1 m (c-d) Fractured - dominated fabric (type 1) @ 2364.8 m & (type 2), grain-matrix are fracture for pore spaces @ depth of 1882.2 m.

[2] in pore size volume and pore throat in each of the different fabric domain types. It also shows that most abundant fabric domain type will be responsible to control the entire fluid flow dynamic at a microscopic scale [22] and thus their efficiency and responses of the rock to stress, pressure, and irreducible water saturation [21]. Each of the mapped fabric domain types is magnified at 500 μm (**Figure 8**) for detail petrographic descriptions, and the summaries are presented in **Table 1**.

Table 1 present fabric domain-type compositional characteristics as studied at magnified 500-μm resolution. The percentage composition of each of the physical textural characteristics is given in the table. It is observable within the table instances of subtypes 1 and 2. These represent a subfabric domain of same fabric-type domain and thus will vary their petrophysical parameters. In both instances, the type 1 represent the domain with lesser matrix composition (<10%), while the type 2 is above 10%. These differences required as already known from the literature that matrix content fill intergranular pore space and will potentially have impact on pore volume distribution, ability to transmit fluid and detainments of irreducible water saturation within the domain of higher matrix composition. It is clear that a particular lithofacies could exhibit a multiple or more than one fabric-type domain(s).

1.3 Fabric domain and porosity distribution

Pore volume is a critical physical parameter that changes and varies within a short distance or space in a porous medium, associated in this context with the different and variable fabric domain type present side-by-side. In order to relate fabric domain type to pore size or porosity distribution, it is important to determine which among the fabric domain exhibits abundant pore space volume within reservoir rock. It is logical that such fabric domain type would have a significant impact on fluid dynamic and retained irreducible water saturation at microscale. Each fabric domain types are expected to have different pore space volume or porosity, and thus pore size and pore throats [21]. From the five (5) sandstone lithofacies where the fabric domain types were mapped (**Figures 2–6**) and summarized in **Table 1**, pore volume or porosity in each of the domain types is quantified and compared with similar domain (in either subclass 1 or 2) in another sandstone lithofacie to infer porosity range that can be associated with each of the domain type. In doing so, predominantly considering the most impactful fabric domain (positively or negatively), one can predict expected macroscopic behavior of the reservoir rock in terms of its irreducible water saturation, potential production rate, and total reservoir efficiency.

Generally, in the clasts-support fabric domain across the three (3) different sandstone lithofacies (**Figure 8a–c**), porosity distribution varies in **Figure 8a** between 10% and 32%, 20% and 34% in **Figure 8b,** and 21–25% in **Figure 8c,** respectively. Porosity in this fabric domain for the subtype 1 is mainly intergranular porosities that vary from 25 to 34%, because it shows moderate- to well-sorted grain sorting, slight compaction, grain-contacts that are point, long, and moderate in concavo-convex, with matrix content less than 10% composition. While in the subclass type 2, characterized by the moderate high-matrix content above 10% obliterating the intergranular pore space, porosity varies from 10 to 21%. This domain will develop smaller pore sizes and pore throats, and will retain high irreducible water saturation, and subtype 1 will develop moderate- to high-conducive pathways for fluid flow due to their larger pore spaces, pore throat, and pore volume potentials [21].

In the random fabric domain (**Figure 8a–d**) characterized with abundance of float and point grain contact, minor to moderate compaction with variable percentage of matrix content, porosity distribution in **Figure 8a** is 11–27%, and in **Figure 7b** is between 21% and 33%, also in **Figure 8c** the porosity varies from 18 to 33%, and from 21 to 24% in **Figure 8d** within the four (4) different sandstone lithofacies. This is associated with the favorable physical rock parameters that enhance the development of potentially more larger pore sizes, pore throats with less matrix composition to retain irreducible water saturation in them, while segment with low porosity values are associated with the slight increase in their matrix composition within this domain as visible in above-discussed domain.

Matrix-supported domain is the most prevalent domain type across the five (5) different lithofacies, because influx of sediments during deposition cannot be controlled or restricted. Porosity distribution within the domain is generally lower compared with aforesaid in the group (**Figure 8a–e**). Intergranular pore spaces are obliterated and reduced by variation in their matrix composition within this domain. Porosity within the domain varies in order of 10–28%, 17–20%, 15–22%, 16–24%, 7–27%, respectively (**Figure 8a–e**). Lower porosity values in this domain that are between 7% and 16% are from the subdomain that exhibit matrix content above 40%, while those that have shown porosity that varies from 20 to 28% are subdomain with matrix composition above 40% both characterized as subdomain types 1 and 2, respectively. Variations in composition of the matrix filling intergranular pore spaces determine distribution of pore volume present within the domain types.

In the laminar domain associated with laminated sandstone lithofacie, fine rock fragments somewhat align forming a visible lamination at microscale. In general, porosity varies in order of 18%–28%, 17%–18%, and 16%–28%, respectively (**Figure 7b–d**). Regions characterized with moderately low matrix composition reveals porosity values between 18% and 28%, and the grains aligning resulted into the development of favorable grain packing and contacts such as the point, and long and moderate concavo-convex that enhances moderate occurrence of intergranular pore spaces within the domain, while that with a moderate increase in matrix content exhibits porosity variation from 16 to 17%. In this domain considering their nearness in porosity, it could be considered as single domain type with no subdomain types despite their variations in the matrix composition. Other textural characteristics are presented in **Table 1**.

The fractured-dominated domain is visible in sandstone lithofacie associated with the occurrence of matrix-fracture porosity [23, 24] during rapid sedimentation [25] probably occurring during a regime of an abnormally high pore-pressure that exceeded hydrostatic pressure at a given depth [26] in the delta. Other textural characteristics (**Table 1**) contribute to development variable occurrence and distribution of porosity within the domain. Porosity distribution in the domain is in order of 22–33%, 18–26%, 22–33%, and 19–29% (**Figure 8b–e**) across studied sandstone lithofacies.

It has indicated that the clasts-supported and random domain types as the most prevalent fabric in reservoir sandstone associated with a high percentage composition of intergranular porosities or pore volumes considering favorable fabric physical parameters such low matrix content, grain sorting, packing/contacts, and moderate compaction contributing to abundance of intergranular porosities. The laminar and matrix-supported domain exhibits next moderate porosities as a result of a high matrix composition filling intergranular pore spaces. At a point, these variable domain types exhibit close to similar volume of porosity [2] within them, but varies at most instance. The fabric domain with the highest pore volume is the fracture-dominated domain that exhibits high occurrence of matrix-fracture pores associated with overpressure at depth as a result of a rapid sedimentation in some sedimentary basin. Furthermore, a lithofacie that exhibit poor intergranular pore-space distribution, termed poorly porous, may experience high permeability values and sudden decline in reservoir pressure due to the occurrence and abundance of the fractured-dominated domains that are not fully noted during petrographic description using polarized microscope only. But if such micrographs are scanned as proposed here, it will be deduce that such reservoir behavior was in response to possible abundance of the fracture-dominated domain presence within the lithofacies.

1.4 Fabric domain and pore size distribution

Pore diameter or pore size is another important reservoir parameter as the width of pores where hydrocarbon is stored. Pore sizes and their interconnectivity

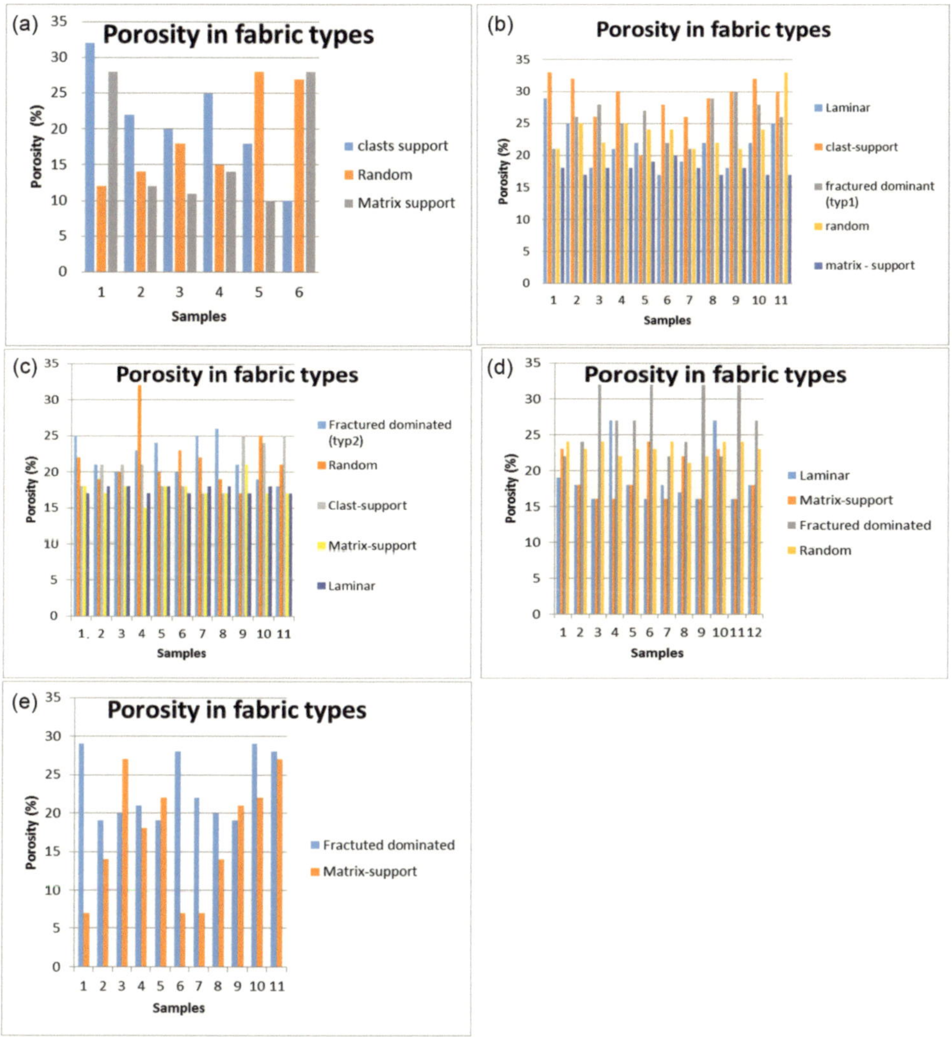

Figure 9.
Porosity distribution in all fabric types within (a) in massive coarse grained sandstone (b) in burrowed medium grained sandstone (c) in faintly laminated fine grained sandstone (d) in parallel laminated friable fine grained sandstone(e) massive very fine grained sandstone.

determine the rate of fluid transmitivity (permeability) across the fabric as their sizes also vary across the fabric domain types. Some domain may exhibit abundance of large pore diameters [21] especially those with moderate matrix composition, mechanical compaction, and favorable physical rock parameter such as grain sorting, grain size, packing/contacts, and moderate diagenetic alteration. For instance, where a particular domain types exhibit increase in aforementioned physical rock parameters, pore size across that domain will be greatly affected through complete occlusion of intergranular pore space or reduction in their pore size diameter across entire rock fabrics. This aspect links fabric domain to variable pore size distribution (**Figure 9**) within mapped fabric domain types to further accentuate that a lithofacie or a single micrograph exhibits multiple fabric domain responsible for the occurrence of multiple hydraulic fluid units (HFU's) in a reservoir [22]. However, it is visible during characterization for hydraulic fluid units (HFU's), where similar in range porosity values alongside their corresponding permeability values from the different domains in a particular or certain lithofacies are plotted or grouped as in reservoir quality index versus porosity in z-direction (RQI Vs. ϕz) to form a Flow Zones Indicator (FZI) unit from different lithofacie successions within a well. The fabric domain types close to similar reservoir quality index ranges in spite of existing in different lithofacies within a succession are plotted as single Flow Zones Indicator

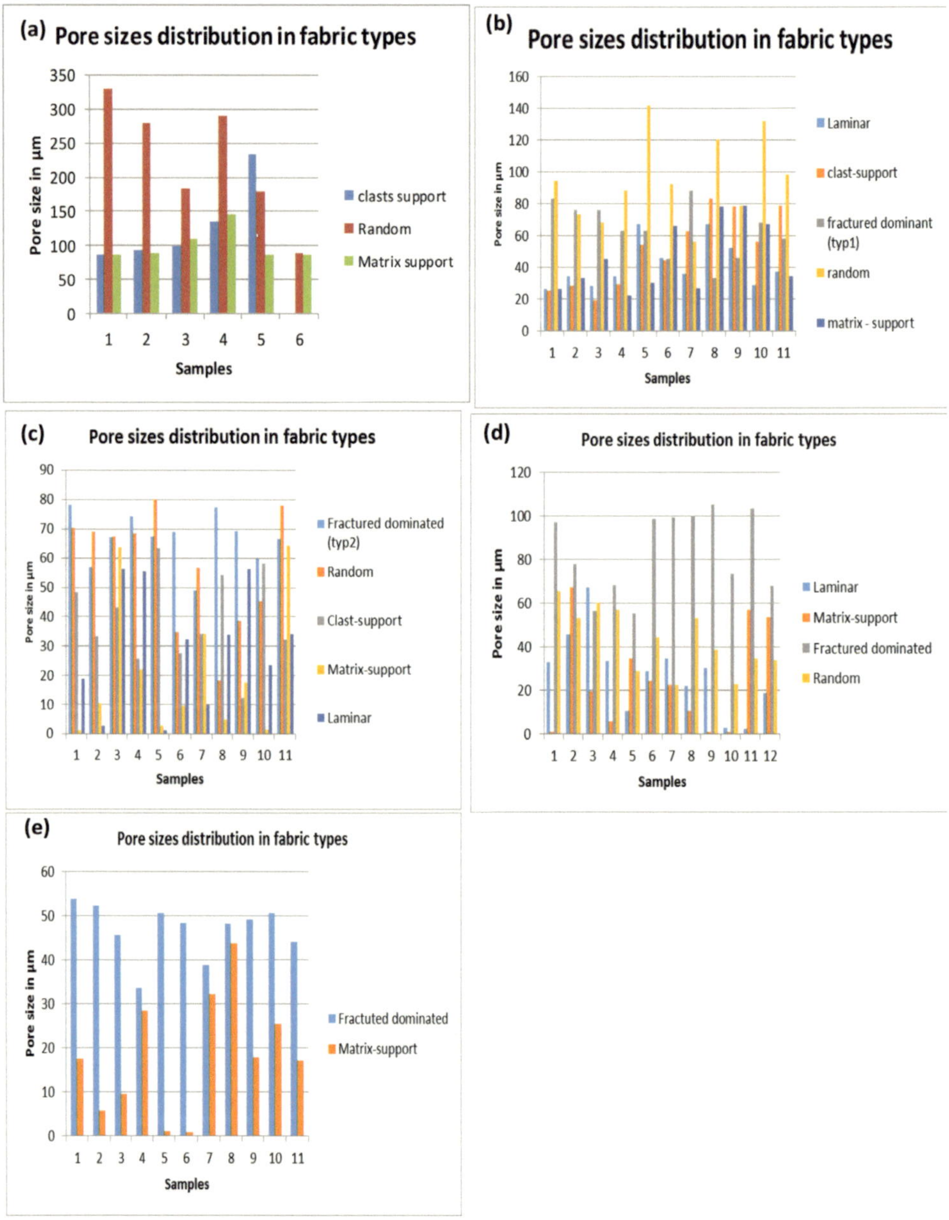

Figure 10.
Pore sizes distribution in fabric types a) in massive coarse grained sandstone (b) in burrowed medium grained sandstone (c) in faintly laminated fine grained sandstone (d) in parallel laminated friable fine grained sandstone(e) massive very fine grained sandstone.

(FZI) units that translate to form high fluid pathways for fluid conductivity within a well bore. Thus, the most dominant and abundant fabric domain types will control the Flow Zones Indicator (FZI) r-squared (r^2) factor values [22]. This interprets that increase or nearness to 1 in r-squared (r^2) factor value is a function of the dominant fabric domain type (s) within investigated lithofacie succession. For an example, when an r-squared (r^2) factor value is 0.9, it indicates that there are more abundance of positive and favorable domain (s) that exhibits larger pore sizes and pore volume with favorable physical rock parameters such as grain size, mechanical compaction, grain contacts; float, long, point moderate in concavo-convex, less in matrix composition, and diagenetic features within their domain. And in an instance with more abundance of the opposite physical rock parameters in a particular domain, the Flow Zones Indicator (FZI) r-squared (r^2) factor value will decrease far from 1 and closer to zero depending on domination of the domain(s).

In **Figure 9a–c**, the clasts-supported domain-type pore size values vary from 62 μm to 248 μm within region of minimum matrix composition, compaction, and diagenetic alteration. In the region with moderately high matrix content, pore size diameter varies between 8 μm and 81 μm within the domain. This domain exhibits the moderate large pore size diameters within it. The regions with large pore sizes presumably will also have large pore throats and good interconnectivity for fluid conductivity less potential to retain irreducible water saturation, while region with a moderate increase in matrix content filling intergranular pore spaces experiences reduction in pore size diameter and certainly will exhibit smaller pore sizes and pore throats, and will retain more irreducible water saturation as result of the matrix composition.

In the random-supported domain (**Figure 10a–d**), pore sizes also vary between 10 μm and 58 μm for the region with small pore sizes, as region with large pore sizes exhibit pore diameter ranging from 65 μm to 330 μm. The domain exhibits the maximum pore size diameter or oversized pores as compared to the clasts-supported domain. This is ascribed to the abundance of intergranular pore space orchestrated by favorable grain sorting and grain packing, and float, point, and long with less occurrence of concavo-convex and sutured contact as result of mechanical compaction due to overburden. The region with an abundant of clasts-to-clasts intergranular pore space with lesser matrix contents and favorable physical rock properties will show larger pore attributes (pore sizes and pore throat) and potentially will retain slighter irreducible water saturation. Conversely, regions with moderate high matrix content potentially characterize with smaller pore attributes and high possibility of retention of irreducible water saturation as a result of higher matrix composition.

In the matrix-supported domain, generally matrix compositions are moderately very high across the entire domain (**Figure 10a–e**), as smaller pore diameters vary between 8 μm and 80 μm, while the larger pore sizes range from 81 μm to 140 μm as their maximum diameters within the domain. In general, this domain exhibits more of medium to smaller pore size and pore throats, because the entire domain is a matrix supported. It will show a high affinity to retain irreducible water saturation within it, and their abundance within a lithofacies will greatly reduce or influence their ability to enhance fluid transmitivity (permeability) and required intergranular pore interconnectivity in fabrics.

The fracture-dominated domain (**Figure 10b–e**) also reveals occurrence of smaller and large pore diameters. In this domain, the large pore sizes vary from 52 μm to 105 μm and those with small pore diameters vary between 25 μm and 49 μm, respectively. In this domain, their pores are extensive and dominant. Thus, this will have an enormous influence on mass fluid flow and conductivity. The moderate smaller diameter pores are characterized with high matrix content termed subclass type 2.

Also in the laminar domain (**Figure 10b–d**), the smaller pore diameter ranges from 0.5 μm to 25 μm, while the large pore diameters vary between 58 μm and 66 μm, respectively. It is therefore logical that the most predominant fabric domain types in a lithofacie control occurrence or development of certain pore sizes and pore volume. Variations in matrix composition and diagenetic features play major roles in pore volume and pore size variation and distributions within same fabric domain type. According to [Archie [2]] pore size distribution does not necessarily define the type of rock, for actually several types of rock may have essentially the same pore size distribution and thus in this context for the different fabric domain types. Therefore, pore size and porosity distribution will vary or comparable within the variable fabric domain types and are responsible for occurrence of multiple hydraulic fluid units (HFU's) and anomalous porosity-permeability relationship in sandstone reservoir rock.

A variation in the fabric shows that on a particular lithofacie could exhibits variable fabric domains spatially and temporally. And also at any given point, one dominant fabric domain could be determined over other domain type (s). Therefore, this proposed concept of a fabric domain needs to be incorporated in detail sedimentological analysis of reservoir rocks. The concept has shown existence of a multiple fabric domains in a lithofacie thin section or micrographs and thus exhibits variation in both their pore size and porosity distribution within the different fabric domain types. These that no fabric of a sedimentary rock is entirely homogeneous thus will have potential impact on pore attributes distribution (pore volume, pore size, pore structure, and pore throat) and irreducible water saturation distribution that determines the distribution of hydraulic fluid unit (HFU's) and entire reservoir fluid dynamics behaviors at a macroscopic scale.

However, there are no long-range processes existing to align the internal rock fabrics as pore volume distribution has dominant influence upon the macroscopic phase separation of mutually immiscible fluids [20]. The proposed concept on existence of multiple fabric domains is visible as pore volume distribution (porosity) and pore sizes within them differ and at some point closer in ranges. Understanding a characterized porous geologic (highly porosity), but low or poor fluid flow transmitivity (permeability) using concept of fabric domain, it could occur that abundance of non-similar domain (differs pore attributes) networks has possibly out numbers the positively similar domain that will form radial intergranular network connectivity for flawless fluid conductivity in lithofacie. Therefore, when investigating, describing variability, or predicting a reservoir macroscopic behavior, the type of heterogeneity should be defined in terms of fabric characteristic and the presence or absence of certain dominant fabric domain type(s).

2. Conclusions

The goal of rock fabric characterization is to describe the spatial and geometric distribution of pore attributes as they impact on petrophysical parameter variations such as porosity, permeability, and saturation. The new insights propose a systematic fabric characterization for siliciclastic sandstone thereby maintaining existing fabric classifications by earlier sedimentologists through the combinations of both digital petrographic scanning and thin-section description to view entirely a lithofacie fabric, such as to examine and visualize potential multiple fabrics instead of a single fabric classification and description as earlier sedimentologist outline in many books.

Fabric domain mapping is achievable through scanning micrographs into high-resolution images at 5-mm scale and carefully visualizes all the different grain assemblages or patterns with reference to their nature of their grain packing, contacts, sorting, orientation, matrix distribution, mechanical compaction, and intergranular pore-spaces into fabric-type domain type (s). It indicates that a single thin micrograph is not ideal to fully a representation of lithofacies rock fabric, and that there are multiple fabric domains existing within a single lithofacies fabric. Five fabric domain types were mapped into clasts-supported, matrix-supported, random, and laminar and fracture-dominated domains some with subclass types 1 and 2. The subclass types 1 and 2 are distinguishable with respect to their individual matrix compositions filling intergranular pore space as in $<10\%$ or $> 10\%$ respectively across the domain types.

The concept of fabric domain types has shown existence of a multiple fabric domain in a lithofacie and pore volume distribution (porosity) and pore sizes within them differ and at some point are closer in ranges. However, a lithofacie that exhibits poor intergranular pore-space distribution, termed poorly porous, may experience high permeability values and sudden decline in reservoir pressure due to the occurrence and abundance of the fracture-dominated domains or oversize pore size within the different domain types that are not fully noted during petrographic description using a polarized microscope only. But if such micrographs are scanned as propose here, it will be understood from the reservoir behavior that it was a response due to the abundance of the fracture-dominated domain or an oversize pores presence within the lithofacies. Therefore, when investigating, describing variability, or predicting a reservoir macroscopic behavior, the type of heterogeneity should be defined in terms of the fabric characteristic and the presence or absence of certain dominant fabric domain types needed to be considered critically in sedimentary rock characterization.

Author details

Ishaq Yusuf[1]* and Eswaran Padmanabhan[2]

1 Faculty of Applied Sciences and Technology, Department of Geology and Mining, Ibrahim Badamasi Babangida University, Lapai, Niger State, Nigeria

2 Faculty of Petroleum Engineering and Geosciences, Department of Petroleum Geosciences, Universiti Technologi PETRONAS, Perak, Malaysia

*Address all correspondence to: ishaq.yusuff@gmail.com

References

[1] Bennett RH, Hulbert MH. Concepts of clay fabric. In: Clay Microstructure. Springer; 1986. pp. 47-103. DOI: 10.1007/978-94-009-4684-2

[2] Archie GE. Introduction to petrophysics of reservoir rocks. AAPG Bulletin. 1950;**34**(5):943-961. DOI: 10.1306/3D933F62-16B1-11D7-8645000102C1865D

[3] Choquette PW, Pray LC. Geologic nomenclature and classification of porosity in sedimentary carbonates. AAPG Bulletin. 1970;**54**(2):207-250. DOI: 10.1306/5D25C98B-16C1-11D7-8645000102C1865D

[4] Lucia FJ. Rock-fabric/petrophysical classification of carbonate pore space for reservoir characterization. AAPG Bulletin. 1995;**79**(9):1275-1300. DOI: 10.1306/7834D4A4-1721-11D7-8645000102C1865D

[5] Tovey N, Sokolov V. Quantitative SEM [scanning electron microscopic] methods for soil fabric analysis [Soil matrix structure]. Scanning Electron Microscope (USA). 1981;**1**:536-554. Available from: https://agris.fao.org/agris-search/search.do?recordID=US8236951

[6] Sokolov V, O'Brien NR. A fabric classification of argillaceous rocks, sediments, soils. Applied Clay Science. 1990;**5**(4):353-360. DOI: 10.1016/0169-1317(90)90030-S

[7] Brewer R, Sleeman J. Soil structure and fabric: Their definition and description. Journal of Soil Science. 1960;**11**(1):172-185. DOI: 10.1111/j.1365-2389.1960.tb02213.x

[8] Launeau P, Robin P-Y. Fabric analysis using the intercept method. Tectonophysics. 1996;**267**(1-4):91-119. DOI: 10.1016/S0040-1951(96)00091-1

[9] Pettijohn FJ, Potter PE, Siever R. Sand and Sandstone. Springer Science & Business Media; 2012. Available from: https://books.google.com/books?hl=en&lr=&id=zcTgBwAAQBAJ&oi=fnd&pg=PA1&dq=Sand+and+sandstone&ots=u1Li4Yepnk&sig=zaeHcttTlAOoBT8b4_3aDbDYNbI

[10] Pettijohn FJ. Sedimentary Rocks. Vol. 3. Harper & Row New York; 1975. Available from: https://webapps.unitn.it/Biblioteca/it/Web/EngibankFile/Sedimentary%20rocks.pdf

[11] Tucker ME. Sedimentary Rocks in the Field: A Practical Guide. John Wiley & Sons; 2011. Available from: https://books.google.com/books?hl=en&lr=&id=0C7Xx-s8J4EC&oi=fnd&pg=PT10&dq=Sedimentary+rocks+in+the+field:+a+practical+guide&ots=gl8Cc8FLyf&sig=5M4YeATcKqUpPTj6j-8vNrtGST0

[12] Boggs S Jr, Boggs S. Petrology of Sedimentary Rocks. Cambridge University Press; 2009. DOI: 10.1017/CBO9780511626487

[13] Lagroix F, Borradaile GJ. Magnetic fabric interpretation complicated by inclusions in mafic silicates. Tectonophysics. 2000;**325**(3-4):207-225. DOI: 10.1016/S0040-1951(00)00125-6

[14] Al-Mukhtar M et al. The fabric of a clay soil under controlled mechanical and hydraulic stress states. Applied Clay Science. 1996;**11**(2-4):99-115. DOI: 10.1016/S0169-1317(96)00023-3

[15] Bryant W et al. The fabric of consolidating clayey sediment column,

ODP Site 697. In: Naval Oceanographic And Atmospheric Research Lab Stennis Space Center Ms. 1990. DOI: 10.2973/odp.proc.sr.113.174.1990

[16] Moon C, Hurst C. Fabric of muds and shales: An overview. Geological Society, London, Special Publications. 1984;**15**(1):579-593. DOI: 10.1144/GSL.SP.1984.015.01.36

[17] Bennett RH, Bryant WR, Keller GH. Clay fabric of selected submarine sediments; fundamental properties and models. Journal of Sedimentary Research. 1981;**51**(1):217-232. DOI: 10.1306/212F7C52-2B24-11D7-8648000102C1865D

[18] Bennett RH, Bryant WR, Keller GH. Clay fabric and geotechnical properties of selected submarine sediment cores from the Mississippi Delta. Unknown. 1977. Available from: https://ui.adsabs.harvard.edu/abs/1977cfgp.rept.....B/abstract

[19] Bennett RH. Clay Fabric and Geotechnical Properties of Selected Submarine Sediment Cores From the Mississippi Delta. Texas A&M University; 1976. Available from: https://www.proquest.com/openview/97c5d9d7a1bd2cdbb0b3f73bb7558f95/1.pdf?pq-origsite=gscholar&cbl=18750&diss=y

[20] Prince CM. Textural and diagenetic controlson sandstone permeability. Gulf Coast Association of Geological Societies Transactions. 1999;**49**(1999):42-53. Available from: https://archives.datapages.com/data/gcags/data/049/049001/0042.HTM

[21] Prince CM, Ehrlich R, Anguy Y. Analysis of spatial order in sandstones; II, Grain clusters, packing flaws, and the small-scale structure of sandstones. Journal of Sedimentary Research. 1995;**65**(1a):13-28. DOI: 10.1306/D426800E-2B26-11D7-8648000102C1865D

[22] Yusuf I, Padmanabhan E. Impact of rock fabric on flow unit characteristics in selected reservoir sandstones from West Baram Delta Offshore, Sarawak. Journal of Petroleum Exploration and Production Technology. 2019;**9**(3):2149-2164. DOI: 10.1007/s13202-019-0617-x

[23] Bratton T et al. The nature of naturally fractured reservoirs. Oilfield Review. 2006;**18**(2):4-23. Available from: https://www.researchgate.net/profile/Paul-Gillespie-4/publication/220010431_The_nature_of_naturally_fractured_reservoirs/links/0046352dd250bc0356000000/The-nature-of-naturally-fractured-reservoirs.pdf

[24] Yusuf I, Padmanabhan E. Delineation of reservoir lateral continuity for flow zones in wells using cluster analysis. Case of West Baram Delta, offshore-Sarawak. International Journal of Advanced and Applied Sciences. 2019;**6**(4):20-32. DOI: 10.21833/ijaas.2019.04.003

[25] Wangen M. Generation of overpressure by cementation of pore space in sedimentary rocks. Geophysical Journal International. 2000;**143**(3):608-620. DOI: 10.1046/j.1365-246X.2000.00248.x

[26] Morley CK et al. Relationship between structural style, overpressures, and modern stress, Baram Delta Province, northwest Borneo. Journal of Geophysical Research: Solid Earth. 2008;**113**(B9):1-23. DOI: 10.1029/2007JB005324

Chapter 5

Intra-Plate Dynamics and Active Tectonic Zones of the Indian Plate

Sanjib K. Biswas and Gaurav D. Chauhan

Abstract

The tectonic framework of the Indian Plate started to evolve since the break-up of Gondwanaland in the Late Triassic. It evolved mainly during the time between its separation from the African plate in the Early-Cretaceous and its collision with the Eurasian plate on the north in Late-Middle Eocene and with the Burmese plate in the northeast in Late-Oligocene. Present active tectonic zones, responsible for earthquake generation, were created by the collision pattern and subsequent plate motion. Continued subduction and plate motion due to ridge push and slab pull are responsible for the activation of primordial faults in the inherent structural fabric of the craton depending on the related stress field. Major tectonic zones of the Indian continental plate are related to the collision fronts and the reactivated intra-cratonic faults along the resurgent paleo-sutures between the proto-cratons. Major Tectonic Zones (TZ) are Himalayan TZ, Assam-Arakan TZ, Baluchistan- Karakoram TZ, Andaman-Nicobar TZ, and Stable Continental Region (SCR) earthquake zone. The structure of the continental margins developed during the break-up of Gondwana continental fragments. Western margin evolved during the sequential separation of Africa, Madagascar, and Seychelles since the Late-Triassic to Late Cretaceous time. The Eastern margin structure evolved during the separation of Antarctica in Mid Cretaceous. The orogenic belt circumscribing the northern margin of Indian plate is highly tectonised as the subduction of the plate continues due to northerly push from the Carlsberg Ridge in the SW and slab-pull towards northeast and east along the orogenic and island arc fronts in the NE. This stress pattern induced an anticlockwise rotatory plate motion. The back thrust from the collision front in the direction opposite to the ridge push put the plate under an overall compressive stress. This stress pattern and the plate motion are responsible for the reactivation of the major intra-cratonic faults. While the tectonised orogenic belts are the zones for earthquake nucleation, the reactivated faults are also the strained mega shear zones across the plate for earthquake generation in SCR. These faults trending WNW-ESE are apparently the transform faults that extend across the continent from Carlsberg ridge in the west to the collision zones in the northeast. As such, they are described here as the 'trans-continental transform faults'. Three such major fault zones from north to south are (i) North Kathiawar fault - Great Boundary fault (along the Aravalli belt) zone, (ii) South Saurashtra fault (extension of Narmada fault) – SONATA-Dauki-Naga fault zone, and (iii) Tellichery-Cauvery-Eastern Ghat-T3-Hail Hakalula-Naga thrust zone. All these trans-continental faults, which are mega-shear zones, are traceable from western offshore to the northeastern orogenic belts along mega tectonic lineaments across the continent. The neotectonic movements along these faults, their relative motion, and displacement are the architect of the present geomorphic pattern and shape of the

IntechOpen

Indian craton. The overall compressive stress is responsible for strain build-up within these fault zones and consequent earthquake nucleation. The mid-continental Sonata-Dauki shear zone follows the Central Indian Suture Zone between Bundelkhand Proto Continent (BPC) and Deccan Proto Continent (DPC). With the reactivation of this shear zone, the two proto-cratonic blocks are subjected to relative movement as the plate rotates anticlockwise. The kinematics of these movements and their implications are discussed here with a special reference to the recent 2001 Bhuj earthquake.

Keywords: active tectonic zones, intra-cratonic faults, trans-continental transform faults, Indian Plate, major plate events

1. Introduction

Crustal plates are styled by the intra-plate stress depending on overall plate dynamics, i.e., break up by stretching, drifting by horizontal forces, and collision/ subduction by convergence. The Indian plate is no exception to this. The structural framework of the Indian plate evolved since its break up from the African plate in Late Jurassic, subsequent northward drift and final collision with the Eurasian plate on the north in Middle Eocene and with Indo-Sinian plate on the northeast in Late Oligocene [1, 2]. Geodynamics of the plate created internal stress activating faults in the pre-existing structural fabric of the Precambrian-Archaean shield. During the break-up stage, when the Indo-African plate was undergoing far-field crustal distension, the intra-cratonic rift basins were formed in Late Jurassic-Early Cretaceous time. In Late Cretaceous post break up crustal rebound and slab-pull towards the north caused trailing edge uplift that aborted the rifting followed by basin uplifts. Drifting motion induced divergent trans-tensional stress on the reactivated faults. Collision and post-collision continued subduction generated compressive stress over the entire plate. This resulted in inversion of the rifted structures. In this paper, we discuss the development of the active tectonic zones (TZ; **Figure 1**) due to varying plate motion during different tectonic set-up and present neotectonic inversion stage.

2. Major plate events and related dynamics

2.1 Pre-break up stage

Before the initial break up in the Permo-Triassic pre-breakup stage, the Eastern Gondwana mega-plate was stretched by far-field crustal distension when the intra-cratonic rifted basins of Gondwana were formed. In the Indian craton, the rifting occurred mostly in the eastern part of the craton (**Figure 2**) as the extensional stress developed mostly between India and Australo-Antarctican plates [3].

2.2 Break up stage

The first break up between Africa and India took place in Late Jurassic and rifting was completed in Early Cretaceous with the separation of Africa and Madagascar-India. As a result, the related intra-cratonic rifting mostly happened in the western pericratonic region of the Indian plate [3]. This was followed by the early Late

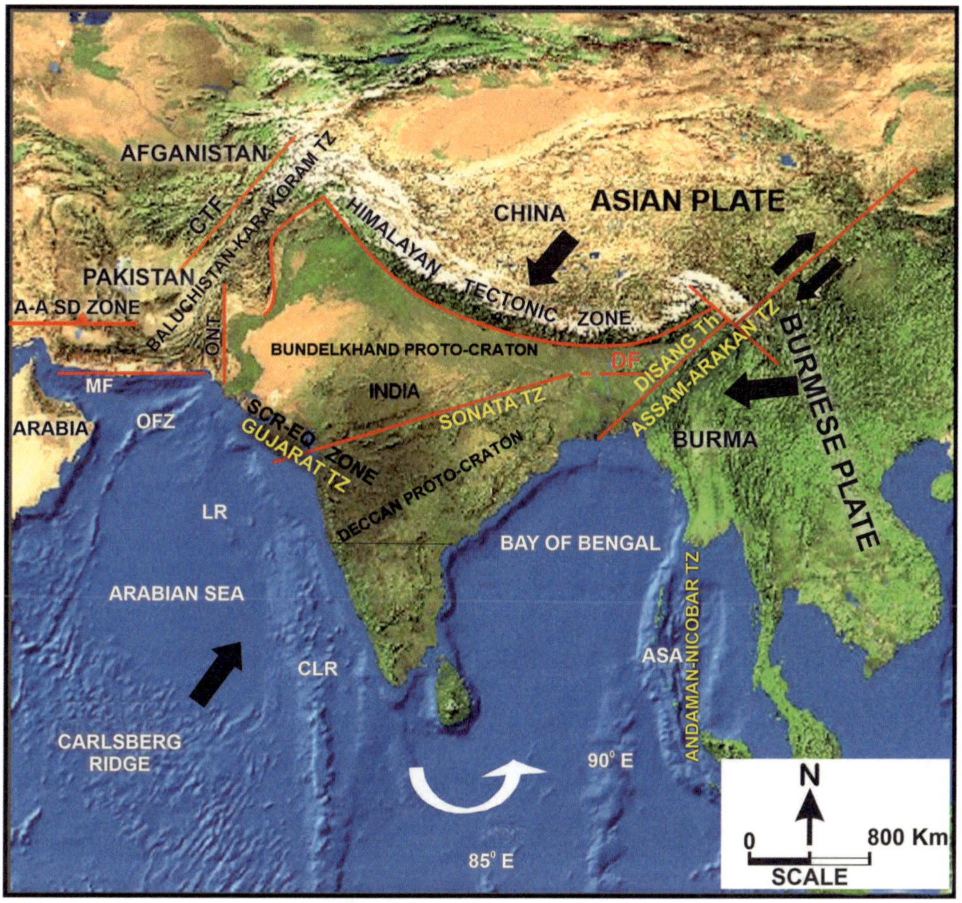

Figure 1.
*Present position of the Indian plate with major active tectonic zones (**TZ**). Black arrows indicate the direction of movement of plates and prevailing stress. The white arrow indicates the direction of plate rotation. Red lines mark major fault/thrust zones. (key:**A-A SD zone**- Afgan-Arabian Ocean subduction zone, **CTF**- Chaman transcurrent fault, **MF**-Makran fault, **ONT**- Ornach Nai Transcurrent fault, **OFZ**- Owen fracture zone, **LR**- Laxmi ridge, **CLR**- Chagos-Laccadive ridge, **DF**- Dauki fault, **ASA**- Andaman-Sumatran arc, **SCR-EQ zone**- stable continental region earthquake zone, **TZ**- tectonic zone, **Th**- thrust).*

Cretaceous break up of India and Australia-Antarctica in the eastern part. Rifting of eastern pericratonic basins and a few Upper Gondwana intracratonic basins took place during this time.

2.3 Rift-drift transition

Rift drift transition occurred in the latest Cretaceous-Early Paleocene time, marked by a widespread unconformity in depositional sequences of both eastern and western pericratonic rift basins. This was a period of stress release and trailing edge uplift of the Indian plate due to the slab-pull from the Tethyan trench. This uplift is responsible for the aborting of the rifts and large-scale upthrusts along primordial faults boosting horst-graben structures along the evolved passive margins where the continents split.

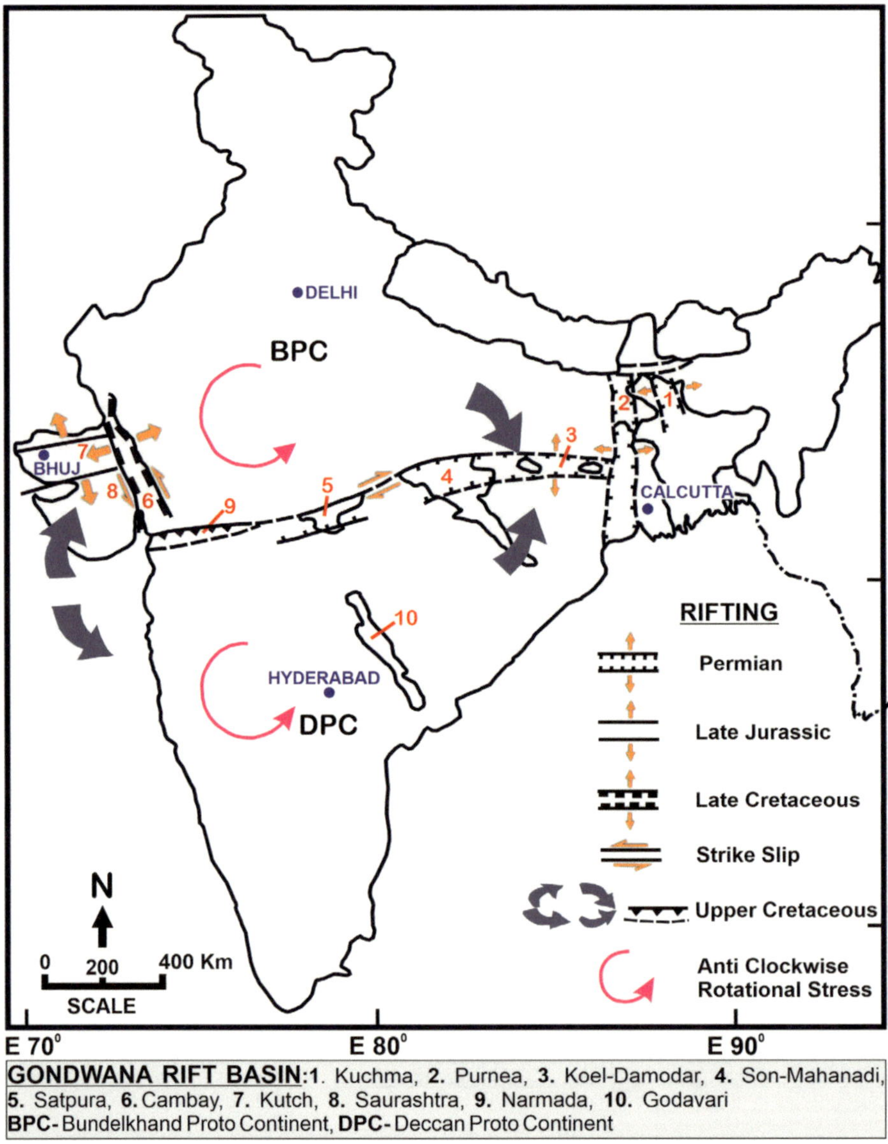

Figure 2.
Map showing major Gondwana and Mesozoic-tertiary rift basins of India and mid-continental NSG. Arrows indicate intra-plate dynamics at different periods.

2.4 Drift stage

As the Indian plate drifted northward with anticlockwise rotation along a curved path (**Figure 3**), the reactivated rift-faults were subjected to horizontal stress, inducing transtensional movements within the pericratonic rift basins. This is evident by the breaking of upthrust-related drape folds along the tilted-up edges of the uplifts (horsts) into small sub-order folds. The best example is seen in the structure of Kutch uplifts (**Figure 4**).

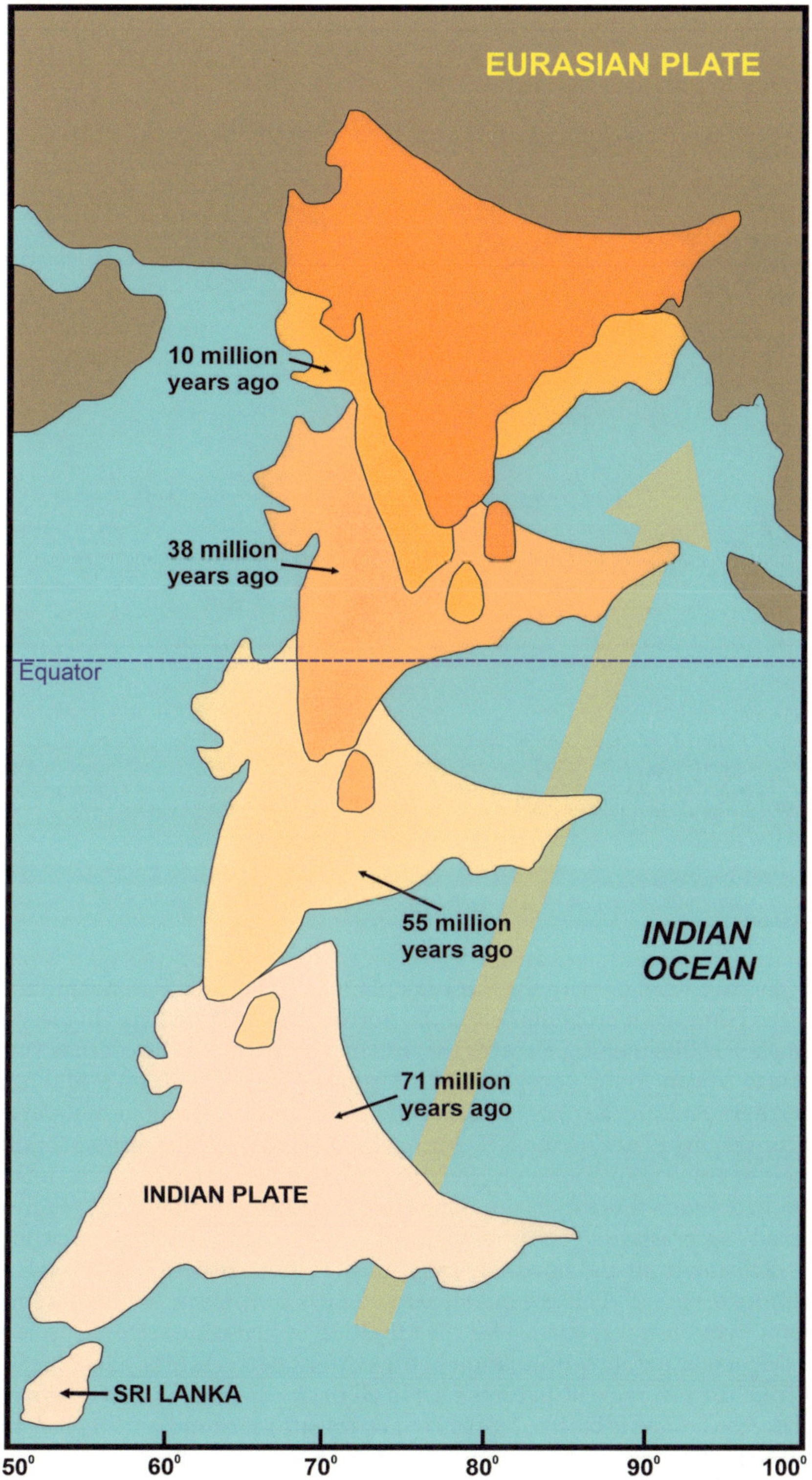

Figure 3.
Path of post-cretaceous drifting of the Indian plate.

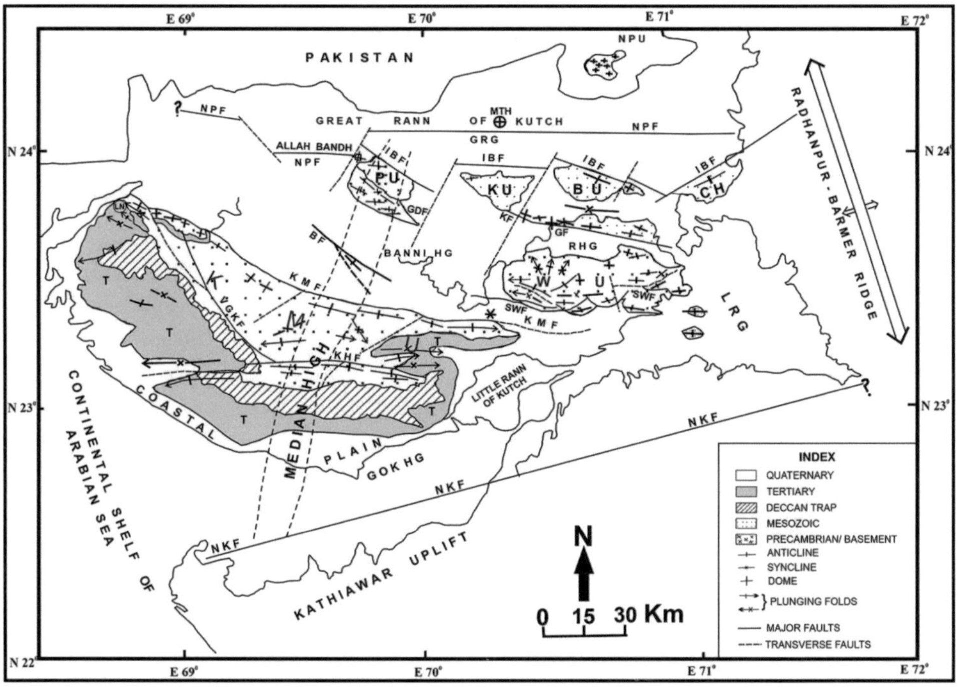

Figure 4.
Tectonic map of Kutch rift basin, the zone of intensive seismic activity in SCR. The map shows major fault-bound tilted uplifts and intervening grabens/half-grabens. Evidence of strike-slip movement is indicated by the breakup of marginal drape fold over the tilted-up edges of the horst (indicated by fold symbols) and left step over of the KMF as SWF towards the east with en echelon shift of Kutch mainland and Wagad uplifts.

2.5 Collision stage

India collided head on with the Eurasian plate in late Middle Eocene initiating uplift of the Himalayan orogenic belt in the north (**Figure 3**). In Late Oligocene Burmese plate collided obliquely with the Indian plate at its northeastern corner giving rise to Assam-Arakan orogenic belt (**Figures 1** and **5**). This stage is continuing as the northern and northeastern edges of the plate are continuing to subduct under the two converging plates. In the NE, due to oblique collision, the plates are still under the process of convergence with progressive southwestward closing of the intervening NE Indian remnant ocean basin, Bengal basin and Bay of Bengal [4] (**Figure 5**). Widespread compressive stress developed in the plate due to southerly backthrust from the collision front and northerly ridge-push from Carlsberg Ridge as the Arabian Sea continues to spread. This compressive regimen is continuing in the present neotectonic cycle. It is responsible for the initiation of inversion tectonic cycle in Eocene-Oligocene and its continuation in the current neotectonic cycle. This is quite evident from the inversion structures seen in all the intra-cratonic basins as well as in the offshore pericratonic basins. Morphotectonic features of India with predominance of the first-order topography also indicate active neo-tectonic cycle dominated by compressive stress.

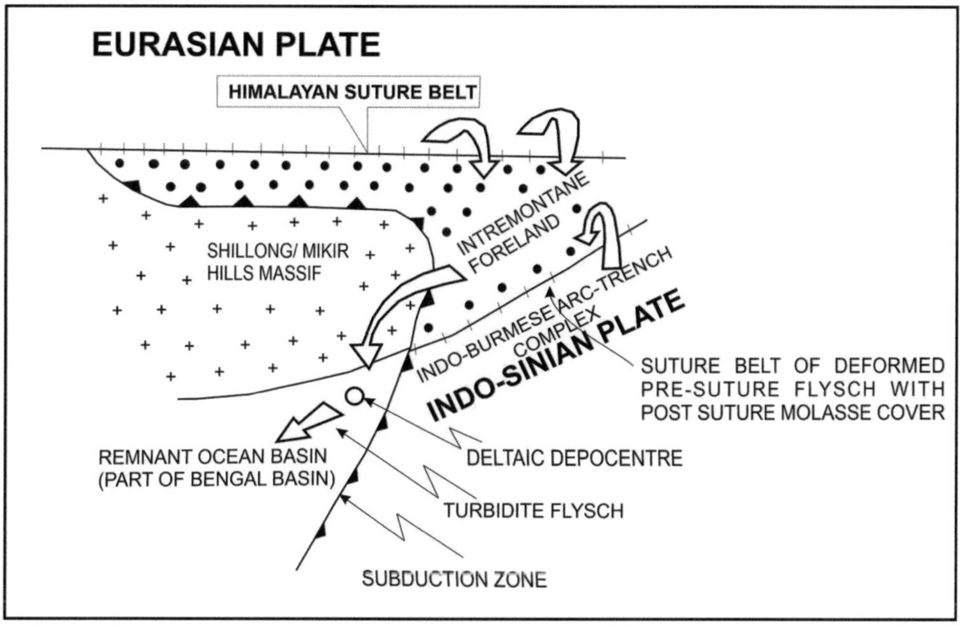

Figure 5.
Plate collision dynamics of NE India – AATZ: The straight collision of Eurasian plate and oblique collision of Burmese plate at the NE corner of India, sutured part is Naga Schuppen belt, and open, the converging region, is the remnant basin of Bengal & bay of Bengal.

3. Active tectonic zones

Several active tectonic zones (TZ) developed in the Indo-Pak Subcontinent (**Figure 1**) during the movement of the Indian plate through the tectonic stages discussed above. These zones are active under compressive stress. The Baluchistan-Karakoram TZ, Himalayan TZ, and Assam-Arakan TZ are present along the northern subduction front of the plate from west to east. A prominent midcontinental TZ, SONATA (Son-Narmada-Tapti) TZ along the NSG (Narmada-Son geofracture), occurs across the plate along a paleo-suture between northern Bundelkhand proto-continent (BPC) and southern Deccan proto-continent (DPC) (**Figure 2**). The tectonic reactivation is taking place due to differential rotating motion between the two proto-cratons along the SONATA TZ. The western pericratonic region covering parts of Maharashtra and entire Gujarat is another active TZ designated here as Gujarat TZ, as evident from the repeated earthquakes in this stable continental region (SCR). The Andaman-Nicobar Island arc is another active TZ as the oceanic plate of northeastern Indian Ocean (Bay of Bengal) is subducting under the arc. These TZs are briefly outlined below.

3.1 Baluchistan-Karakoram TZ

The Karakoram-Himalayan orogenic belt is the subduction complex along the northwestern periphery of the Indian plate (**Figure 1**). The northernmost projection of the leading edge of the plate in the region of Rawalpindi and Jammu had the first contact with the Eurasian plate. Subsequently, the subduction was affected by northward motion with simultaneous anticlockwise rotation of the plate. Thus, the northwestern part of the leading edge has a transformal relationship with the Afghan craton. The transpressional strike-slip relationship is marked by Chaman transcurrent (CT) and Ornach-Nai transcurrent faults (ONT). The Baluchistan arc marks the subduction complex of the Arabian Sea and the Afghan craton collision [5]. All these faults are presently active making this TZ a prime earthquake-prone zone.

3.2 Himalayan TZ

The Himalayan TZ marks the continent-continent collision zone with the ongoing thrusting of the Tibetan plate over the Indian plate. This zone, therefore, is highly vulnerable to earthquake generation. Epicentres of several disastrous earthquakes are located in this zone. The northward motion of the Indian plate is constrained by this collision front which is building up strain in this zone and also back thrust that is responsible for the compressive stress experienced in the sub-continent.

3.3 Assam-Arakan TZ

This is a zone of oblique collision where two colliding continents are still in the process of convergence with a remnant ocean between them [6]. The Burmese plate collided with the NE corner of the Indian plate near the present syntaxial bend of the mobile belt (**Figures 1** and **5**). The plate continues to rotate towards the northeastern edge of the Indian plate as suturing is progressing southwestward with the extension of the subduction complex of the mobile belt. As such, this zone is tectonically highly active in the present cycle. This is evident from the intense seismic activity and occurrence of many strong earthquakes.

3.4 SCR Earthquake zone (Gujarat TZ & SONATA TZ)

The SCR earthquake zone includes parts of central and western India covering parts of Madhya Pradesh, Maharashtra, and almost the whole of Gujarat (**Figure 1**). The rifted region of Kutch-Cambay, Saurashtra, and Narmada comprises the SCR EQ-zone. The Son-Narmada-Tapti tectonic lineament zone, SONATA, across the Indian shield is a part of this SCR EQ zone. This ENE-WSW trending SONATA zone is defined by Narmada-Son lineaments in the north and Tapti lineament in the south. The zone is reactivated along Precambrian Satpura-Bijawar mobile belt occupying the Central Indian Tectonic Zone (CITZ) [7]. The northern part of the CITZ is the suture zone between the BPC and DPC (Dharwar-Bastar-Singbhum) proto-cratons or sub-plates. The zone consists of a bunch of E-W striking faults parallel to the NSG, reactivated as right-lateral strike-slip faults in the neotectonic cycle. It is affected by NE-SW striking Burhanpur wrench fault [8] with a right-lateral shift. The active seismic zone around Jabalpur, Broach, and Surat is a part of this active tectonic zone. Gujarat TZ includes Saurashtra horst, Kutch, Cambay, and Narmada rifts with active faults. The most vulnerable area of strain build-up for earthquake generation is the Kutch rift (**Figures 4** and **6**).

3.5 Andaman-Nicobar TZ

This is an Island arc, a part of the Sumatran arc which is the collision front of the oceanic plate of the Indian Ocean and the Indochina-Malaysian continental

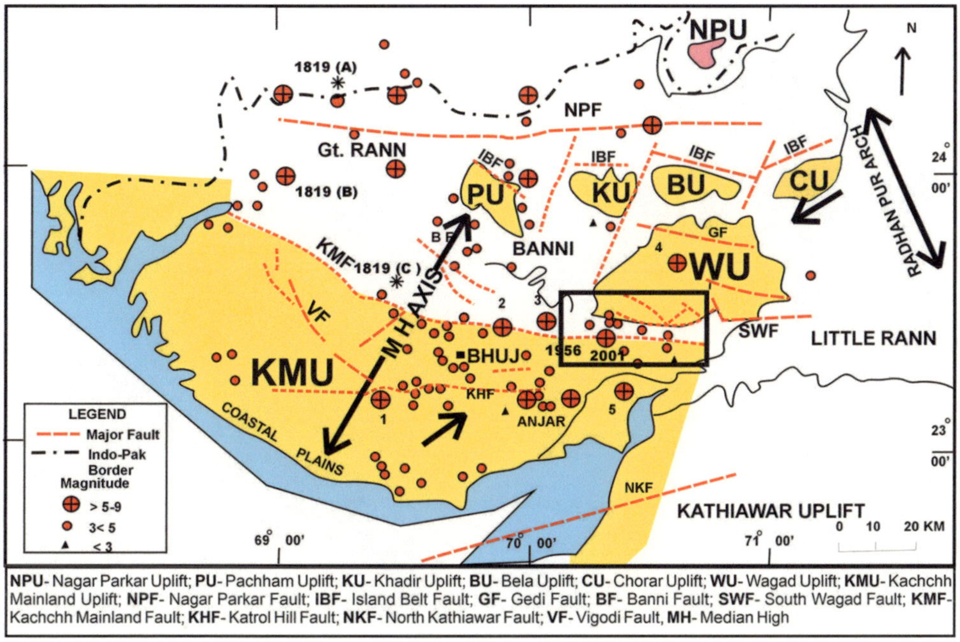

Figure 6.
Tectonic map of Kutch showing major tectonic elements which are major stress barriers for strain build-up. Arrows indicate stress directions. Square marks the critical strain build-up zone in the fault step-over area. Stars and crossed circles within the critical zone are locations of major earthquake epicentres.

plate (**Figure 1**). It extends northward into the Assam-Arakan orogenic belt where the oceanic plate has been consumed as the Burmese plate converged on the Indian plate. This Arc is very active seismic zone where the disastrous Tsunami of 2004 originated.

4. Transcontinental mega-shear zones (MSZ) and plate dynamics

Indian Plate is affected by five main ocean-to-continent transcurrent faults as indicated by the extension of important offshore transform/strike-slip faults across the continent. These are, from north to south, the North Kathiawar-Great Boundary fault, SONATA Zone, Alibag, Vengurla, and Tellichery-East Coast-HHL-Naga Hills faults (**Figure 7**) (HHL: Hail-Hakalula lineament). The trans-continental extension of these faults is traced by strong tectonic lineaments matching with mapped fault/shear zones and important Proterozoic tectonic trends. The matching strikes of North Kathiawar and Great Boundary Fault suggest a continuous trend of crustal shear between Trans Aravalli proto-craton (TAPC) and BPC. These extensive and active fault zones are considered here as mega shear zones (MSZ).

Presently the Indian plate is under compressive stress ([9]; **Figures 1** and 7). The slab-pull from the Andaman trench is causing the anticlockwise rotation of the plate (**Figure 7**). The Indian plate is divided in the middle by the SONATA TZ which is a mega-shear zone (MSZ) reactivated in the present neotectonic cycle as a transcontinental transform fault (**Figure 7**). This MSZ extends from the Carlsberg Ridge to Upper Assam across the continent along NSG connecting the Dauki fault and Naga

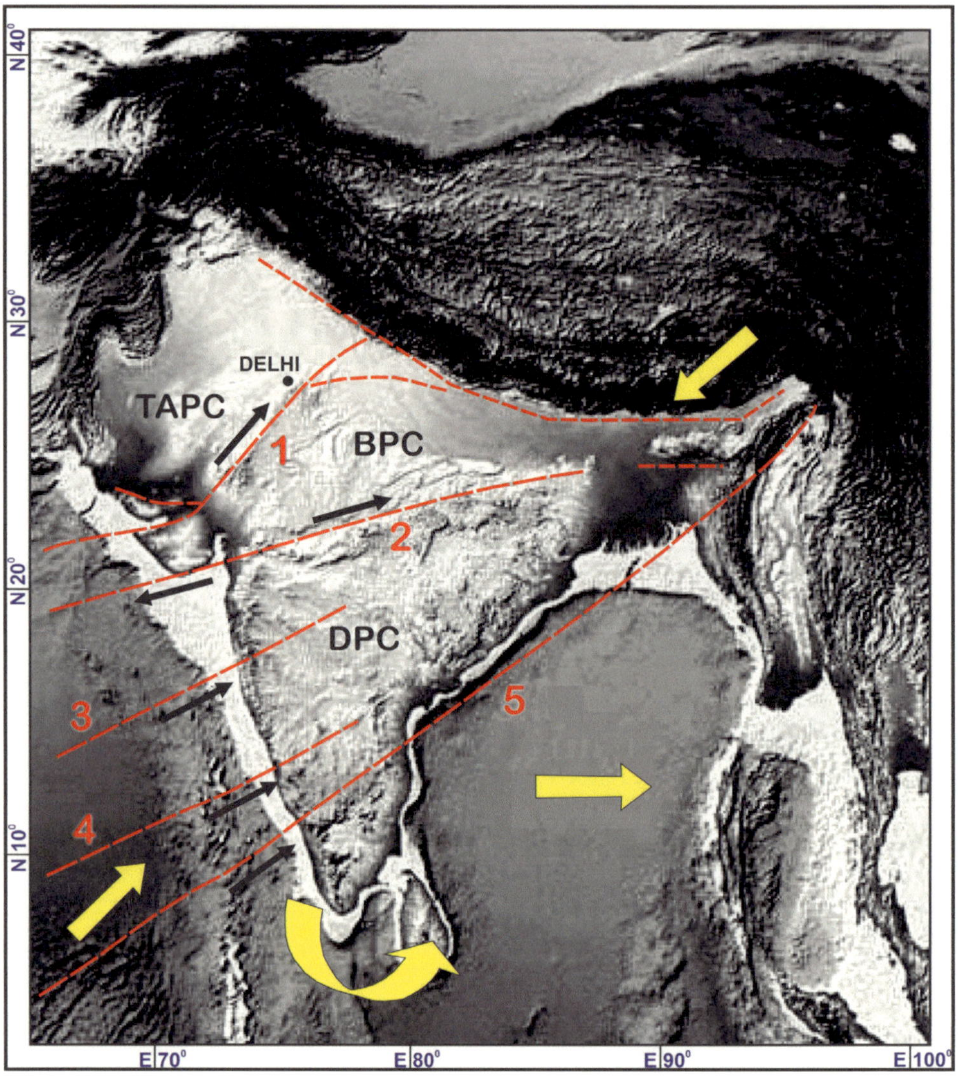

Figure 7.
Red lines mark the major ocean-to-continent transcurrent faults (MSZs, numbered) across the Indian shield: (1) North Kathiawar – Great Boundary Fault, (2) Narmada-Son-Dauki-Naga Fault, (3) Alibag Fault, (4) Vengurla Fault, (5) Tellichery – E. coast-HHL-Naga Thrust. Stress/movement directions are shown by black arrows. Yellow arrows show prevailing regional stress directions following plate movement.

thrust [9]. As a result, the two proto-cratons, BPC & DPC, are rotating with differential motion on either sides of this mid-continental shear zone (**Figure 2**) [3]. The motion of the northern protocraton is constrained by the collision front whereas the southern craton is moving relatively free in response to the anti-clockwise plate motion.

The Deccan sub-plate is affected by another mega-shear zone, the Tellichery fault, extending from Carlsberg Ridge in the offshore to Naga frontal thrust along the Naga Hills in AA TZ. This fault extends across the southern part of DPC through the Palghat gap, Kaveri shear zone, along the east coast (bordering Krishna- Godavari rift basin), and across the Bangladesh-Tripura fore-arc prism following Eastern Ghat Precambrian trend (**Figure 7**). This is defined here as Tellichery-Naga-Hills MSZ.

Between SONATA MSZ and Tellichery-Naga Hills MSZ, two other offshore faults, Alibag and Vengurla faults, occur. These faults also appear to extend across the shield but the lineaments are obscured by the Deccan Trap cover. The relatively free rotation of the Deccan subplate is creating a tensional stress in the region of the Gulf of Cambay and Narmada (**Figure 2**). This is evident from the occurrence of pull-apart basins in this region [3, 9]. At the same time, in the central and eastern parts of this MSZ, transpressional stress is developed (**Figure 2**). This is evident from the uplift of the Gondwana rifts in the central and eastern parts of this MSZ. South of NSG, the three MSZs across the Deccan sub-plate divide the plate into slices which are slipping left-laterally relative to each other from north to south due to rotation of the plate. This progressive left lateral shift from north to south is apparently responsible for the convex outline of the present coastline.

The Tellichery-Naga MSZ is a resurgent shear zone playing an important role in the present-day plate dynamics. The identification of the mega shear extending from the Carlsberg ridge to the Indo-Burmese plate boundary adds a new dimension in the plate kinematics in the northern Indian Ocean as it appears to be a new or evolving transform plate boundary. Between Eastcoast and AA TZ this MSZ passes through an active zone of seismic activity (**Figure 8**) and it matches with the active TT3 and HHL tectonic lineament of Bangladesh [10] and Tripura-Naga Hills [11] respectively. This transform motion and the stress generated by active convergence of Indian and Burmese plates following oblique collision are responsible for the high degree of seismicity of the Assam-Arakan TZ.

The compressive stress due to continuing north and north-northeastward subduction of the Indian plate is responsible for the seismicity of the Himalayan TZ. The Baluchistan-Karakoram TZ (**Figure 1**) is also highly vulnerable to earthquakes. The recent 2005 Baluchistan earthquake is an example. This is caused by different plate motions along the AA-SD TZ in this northwestern border of the plate. The compression related to the continuing northward subduction of the plate along the Karakoram thrust, the transform motion between the Indian and Afghanistan plates along CT and ONT, and subduction of the Arabian Sea oceanic plate below the Afghan plate along the Makran Fault (MF) in AA-SD TZ, west of the transform boundary are causative forces.

In the SCR zone, the highly rifted Gujarat region is the most active seismic zone in peninsular India. The structural inversion of the rifted structures due to present compressive stress is responsible for the repeated generation of the large earthquakes M > 7.0, particularly in the Kutch rift where the confining stress is enhanced by the local structural framework as discussed below. The SONATA zone is another earthquake-prone linear zone. Several major strong earthquakes M ~ 6.0 occurred around Jabalpur in the past including the recent 1997 earthquake [12]. The focal depth of the 1997 Jabalpur earthquake is estimated at 35 km, at the crust–mantle boundary [13]. The dextral strike-slip motion and related kinematics associated with the parallel faults and their conjugate Riedel faults in the SONATA are the cause of repeated rift basin deep crustal earthquakes within this zone as noted in cases of the 1973 Broach and the 1997 Jabalpur earthquakes M > 6.0.

The Latur and Koyna earthquakes are apparently related to the Koyna-Kurduwadi rift (**Figure 9**) inversion with compressional stress [15]. These rifts are apparently related to Alibag MSZ passing south of the SONATA zone. These events are, however, shallow (depth < 10 km) upper crustal earthquakes.

4.1 The 2001 SCR (BHUJ) earthquake

January 26, 2001, Republic day EQ earthquake in Bhuj, Gujarat state, is a world example of a recent high magnitude Mw 7.7 earthquake in SCR. Several disastrous

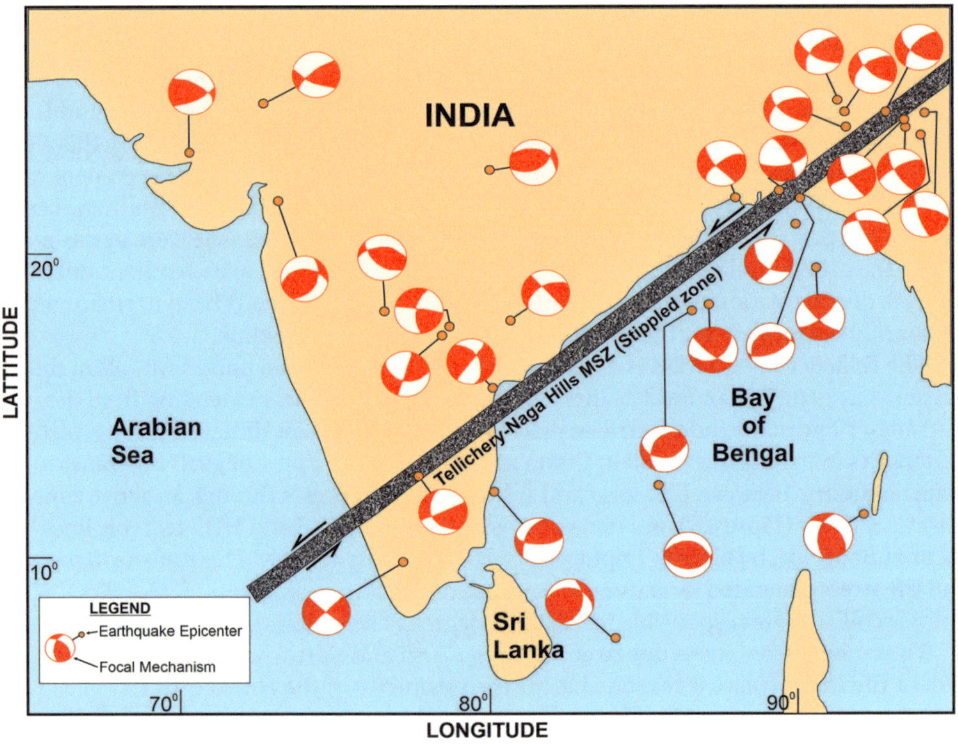

Figure 8.
Map showing the focal mechanism of seismic events along Tellichery-Naga Hills MSZ (stippled zone). (Courtsey: Dr. C. Subrhamanyam, NGRI).

earthquakes occurred in the Kutch rift since ancient times. Strain build-up at the E-W master faults due to intra-plate kinematics is the reason for repeated earthquake generation [16]. The Kutch Mainland Fault (KMF) in the middle of the rift is the main active fault for earthquake generation (**Figure 4**). This fault is currently experiencing dextral transpressional strike-slip movement. Towards the east, the fault tapers off and sidesteps to the left (i.e., shifts to the north) and continues eastward as South Wagad Fault with an approximately 50 km step-over zone (**Figures 4** and **6**).

Intense seismic activity within this step-over zone is indicated by crowding of earthquake epicentres including two major high-intensity earthquakes, the 1956 Anjar, and the 2001Bhuj (**Figure 6**) earthquake. This fault step-over zone is strained by the accumulation of regional compressional stress. Further, the occurrence of massive plutons and geophysical data indicate the presence of a deep-seated igneous body which appears to be syn-rift crustal melt in the deeper crust at 20–40 km depth (**Figure 10**). Seismic tomography study [17] in the Bhuj earthquake epicentre area clearly indicated fluid-filled rock matrix at this depth [18]. The E-W rift ends up against an NW-SE trending basement ridge, the Radhanpur-Barmer arch that separates this rift and the transversely oriented N-S Cambay rift (**Figures 4** and **6**). The easterly horizontal stress along KMF/SWF is constrained by this ridge, which acts as an effective stress barrier. This adds to the strain build-up due to compressive stress within the critical stepover zone. The resistance against the igneous body further

Author details

Sanjib K. Biswas[1] and Gaurav D. Chauhan[2*]

1 Former Director, Keshava Deva Malaviya Institute of Petroleum Exploration (KDMIPE), Oil and Natural Gas Corporation, Dehradun, Uttarakhand, India

2 Department of Earth and Environmental Science, KSKV Kachchh University, Bhuj, Gujarat, India

*Address all correspondence to: gdc.dew@gmail.com

References

[1] Deitz RS, Holden JC. Reconstruction of Pangea, break up and dispersion of continents, Permian to Present. Journal of Geophysical Research. 1970;**75**:4939-4956

[2] Powell CM, Roots SR, Veevers JJ. Pre-break up continental extension in East Gondwanaland and the early opening of the eastern Indian Ocean. Tectonophysics. 1988;**155**:261-283

[3] Biswas SK. A Review on the evolution of rift basins in India during Gondwana with special reference to Western Indian basins and their hydrocarbon prospects. In: Sahni A, Loyal RS, editors. Gondwana Assembly: New issue and Perspectives. New Delhi: Ind. Nat. Sc., Acad; 1999. pp. 53-76

[4] Biswas SK, Bhasin AL, Ram J. Classification of Indian sedimentary basin in the framework of plate tectonics. In: Proc. Second Seminar on Petroliferous Basins of India, Dehradun. Vol. 1. 1993. pp. 1-46

[5] Zaighm NK, Malliick KA. Prospect of hydrocarbon associated with fossil rift structures of southern Indus basin, Pakistan. Bulletin of the American Association of Petroleum Geologists. 2000;**84**:1833-1848

[6] Biswas SK. The Upper Assam inter-orogenic peripheral foreland basin – A special type of peripheral foreland basin in Northeast India – An overview of tectono-sedimentary evolution, Souvenir vol. of National Seminar on Geodynamics, sedimentation and biotic response in the context of India-Asia collision, Mizoram University & Geol. Soc. Ind., 2009. pp. 77-90

[7] Acharyya SK, Roy A. Tectonothermal history of the Central Indian tectonic zone and reactivation of major faults/shear zones. Journal Geological Society of India. 2000;**55**(3):239-256

[8] Ravi S. Thermal and crustal structure of SONATA, a zone of mid-continental rifting in Indian shield. Journal Geological Society of India. 1991;**37**(3):211-220

[9] Biswas SK. Geodynamics of Indian plate and evolution of the Mesozoic-Cenozoic basins. Memoir Geological Society of India. 2008;**74**:247-260

[10] Khan AA, Chouhan RKS. The crustal dynamics and tectonic trends in Bengal basin. Journal of Geodynamics. 1996;**22**:267-286

[11] Khar BM, Ganju JL. Tectonics of Tripura folds - probable mechanics of folding and faulting. Petroleum Asia Journal. 1984;**1**:66-70

[12] Thakur VC, Sushil K, Phillip G. May 22, 1997, Jabalpur earthquake: reactivation of eastwest trending Jabalpur fault in the Narmada-Son lineament zone. Himalayan Geology. 1998;**19**(1):119-126

[13] Bhattacharya SN, Ghosh AK, Suresh G, Baidya PR, Saxena RC. Source parameters of Jabalpur earthquake of May 22, 1997. Current Science. 1997;**73**:855-863

[14] Kale VS, Kulkarni HS, Peshwa VV. Discussion on a geological map of the southern Deccan Traps, India and its structural implications. Journal of Geological Society. 1991;**148**:495-505

[15] Misra DC, Gupta SB, Rao V. Space and time distribution of gravity field in earthquake affected areas

of Maharashtra, India. Memoirs. 1994;**1994**:119-126

[16] Biswas SK, Khatri KN. A geological study of earthquakes in Kutch, Gujarat, India. Journal of Geological Society. 2002;**60**:131-142

[17] Mandal P, Rastogi BK, Satyanaraya HVS, Kousalya M, Vijayraghavan R, Satyamurty C, et al. Characterization of the causative fault system for the 2001 Bhuj earthquake of Mw 7.7. Tectonophysics. 2004;**378**:105-121

[18] Kayal JR, Zhao D, Mishra OP, De R, Singh OP. The 2001 Bhuj earthquake: Tomography evidence for fluids at hypocenter and its implications for rupture nucleation. Geophysical Research Letters. 2002;**29**(24):51-54

[19] Biswas SK. A review of structure and tectonics of Kutch Basin, Western India with special reference to earthquakes. Current Sciences. 2005;**88**(10):1592-1600

Section 2

Aquifers

Chapter 6

Management of Hard Rock Basaltic Aquifer through Aquifer Mapping: A Case Study of Nashik District, Maharashtra, India

Prabhat Jain, Abhay Soni and Rahul Shende

Abstract

In the Maharashtra State of India, Deccan Trap basaltic lava flows are spread over around 82% of the area and form the most prominent aquifer in the entire state. Nasik district occurring in the northern part of Maharashtra also known as Khandesh represents a typical area of Deccan Trap basalt. The storage and transmission capabilities of the basaltic lava flow aquifer are very limited due to the inherent absence of primary pore spaces. These basaltic rocks act as aquifers only when they are weathered, jointed or fractured, thus giving rise to secondary porosity and permeability. Due to wide variations in secondary openings, the potential areas for groundwater are generally localized. In this way, Deccan Trap basalt possesses a unique challenge to aquifer mapping, both spatially and vertically due to its hydrogeological heterogeneity. In the current study, this challenge of aquifer mapping and management in basalt was tackled through a multidisciplinary, multipronged approach involving data integration of various thematic layers viz., geomorphology, soil, drainage, land use-land cover, hydrometeorology, and geophysical techniques etc., as indirect tools and combining it with direct tools such as drilling, well inventory, water level monitoring, groundwater quality checks, and aquifer pumping tests for obtaining reliable results. By following the above methodology, the 3-D aquifer geometry, lithological sections, fence diagrams, aquifer characteristics, yield potentials, and aquifer-wise resources were deciphered. The results showed that the area has two aquifer systems comprising of Aquifer-I, that is. shallow aquifer, which is generally tapped by the dug wells of 8 to 32 m depth with water levels of 1.2 to 15 meters below ground level (m bgl) and yield varies from 10 to 100 m^3/day. Whereas, the Aquifer-II, that is, deeper aquifer is being tapped by bore wells with a depth ranging from 30 to 200 m bgl and a water level from 8 to 55 m bgl. However, their pumping sustainability was limited to 0.5 to 3 hours due to low storage potential resulting in overexploitation. The given aquifer maps indicate that major parts of the area have limited yield (Aquifer-I: between the depth of 10 m - 15 m bgl and Aquifer-II: between the depth of 80 m –140 m bgl). In hard rock areas, especially basaltic aquifers due to their low storage potential, groundwater development is always a challenging task unless it is combined with the management of the resources. Considering the issues plaguing the area, the aquifer management plan encompassing supply and demand-side interventions, and groundwater development has been devised. It is concluded that 139.30 MCM of groundwater resources can be augmented by artificial recharge under supply-side interventions. Whereas the groundwater demand for irrigation can also be reduced by 272 MCM by adopting drip irrigation in 117 sq. km. of sugarcane and 790 sq. km. of onion cultivated areas under the demand-side interventions. The implementation of these measures will minimize the stress on groundwater by bringing down the stage of groundwater development from 88–55% (safe category) in six water-stressed blocks/taluka, whereas the overall stage of groundwater development will be reduced from 58.45% to 40.70%. Thus, the adoption of both supply-side and demand-side interventions interlinked with water budgeting through community participation will provide long-term solutions to combat the overexploitation, water level decline, low storage potential, recurring droughts and other issues of the area and also help in improving socioeconomic conditions of the area.

Keywords: Deccan trap basalt, hydrogeological heterogeneity, aquifer mapping, data integration, 3D aquifer geometry, aquifer characteristics, overexploitation, supply-side/demand-side interventions

1. Introduction

Groundwater is often referred to as an invisible resource, controlled by many natural factors such as physio-climatic conditions, geomorphology, topography, soils and the most important being the geology thereby hampering its sustainable development and management. However, when groundwater is hosted by a hard rock aquifer, its quantifiable attributes become even more unpredictable, and site-specific, thus making it a challenge to devise a suitable management plan. Aquifer mapping is one such tool to effectively manage hard rock aquifers. Aquifer mapping is a process wherein a combination of geological, geophysical, hydrological, and chemical analyses is applied to characterize the quantity, quality, and sustainability of groundwater in aquifers.

The activities include the collection and interpretation of data available from groundwater from various authentic sources. Having plotted the data on maps the data gaps are identified by micro-level data acquisition, that is. through the drilling of bore wells down to the depths of 200 meters and the generation of records of requisite data. Hydrological and hydrometeorological studies for recharge estimation and extensive quality monitoring to assess the potability of groundwater for various uses. Conducting various aquifer performance tests to ensure the yield potential of the aquifer and its sustainability for optimum pumping is the next step followed. Using all these data and data sets various maps have been prepared to draw the 3D geometry of the shallow and deeper aquifers occupying up to 200 m bgl in the area. Both quality and quantity have been assessed to ascertain groundwater availability and safe development (exploitation) in the study area. GIS technology has been used widely for the preparation of various aquifer maps and also to draw the aquifer management plan forming a part of National aquifer mapping (NAQUIM) [1].

In the Maharashtra state of India, Deccan Trap basaltic lava flows are spread over around 82% of the area and form the most prominent aquifer in the entire state. The storage and transmission capabilities of the basaltic lava flow aquifer are very limited due to the inherent absence of primary pore spaces. These basaltic rocks act as aquifers only when they are weathered, jointed, or fractured thus giving rise to secondary porosity and permeability. Due to wide variations in secondary openings, the potential areas for groundwater are generally localized. Thus, Deccan Trap basalt possesses a unique challenge to mapping the aquifer both spatially and vertically due to its hydrogeological inhomogeneity.

2. Materials and methodology

The aquifer mapping study under NAQUIM was taken up by Central Ground Water Board (CGWB) in 3 phases and completed in March 2022. One of the important aspects of aquifer mapping is data gap analysis because it determines the optimum data, which is required to be generated after considering the existing available data as per the aquifer mapping protocols. Without the availability of sufficient data, a true picture of the area cannot be generated [2]. The available data of the exploratory wells, groundwater level monitoring stations, micro-level hydrogeological data acquisition, and groundwater quality monitoring stations of CGWB were compiled and analyzed for adequacy. In addition to these, the data on groundwater monitoring stations and groundwater quality stations of the groundwater survey and development agency (GSDA) was also utilized [3]. After taking into consideration the available data on groundwater exploration, geophysical survey, monitoring, and quality, the data adequacy has been worked out considering the heterogeneity of aquifers in the area.

The thematic layers of geomorphology, soil, drainage, and land use-land cover were not available, hence considered as a data gap. The data gap analysis was done for major parameters namely, exploratory drilling data, micro-level hydrogeological data (acquisition from the key well inventory/establishment), groundwater monitoring data, and groundwater quality data. The data gap is arrived at by considering the requirement and the existence in the grid pattern whose details are as follows -

- Exploratory wells = 14 wells (details available, but not adequate as per the grid-pattern requirement)
- Gaps = 50 wells (instead of 40 wells)
- Micro-level data gap = 148 wells
- Water level monitoring wells = 154 wells (quality data gaps)

Considering the data gaps, feasibility and availability of locations, additional data on exploratory wells, micro-level hydrogeological well inventory, water level monitoring, and groundwater quality was generated (**Table 1**). The drilling data of 61 wells comprising exploratory wells, observation wells, and piezometer was used to prepare the subsurface lithology, hydrogeological cross sections, fence diagram, and aquifer maps. The water level data of 215 monitoring wells of CGWB and GSDA of Aquifer-I, and 55 wells of Aquifer-II of CGWB was utilized to depict the aquifer-wise water level

Exploratory data			**Micro-level HG data acquisaition**			**Gwmonitoring data**			**Gwqualitydata**		
Req.	**Exist.**	**Gap**	**Req**	**Exist.**	**Gap**	**Req.**	**Exist.**	**Gap**	**Req.**	**Exist.**	**Gap**
54	14	50	148	0	148	215	85	154	215	85	154

Table 1.
Data gap analysis.

scenarios during pre-monsoon and post-monsoon seasons. The long-term water level data for the period 2012–2021 was available for 185 monitoring wells out of 215 wells, and it was analyzed to depict the water level trends for Aquifer-I. The groundwater samples collected primarily during the pre-monsoon season from 128 dug wells, and 35 borewells were analyzed for physicochemical constituents *viz*., turbidity, pH, electrical conductivity, total dissolved solids, alkalinity, total hardness, calcium, magnesium, chloride, sulphate, nitrate, fluoride, and iron. The results of the analyses were used to determine the aquifer-wise groundwater quality of the area. The micro-level hydrogeological well inventory was done at 148 locations to have detailed information on the subsurface strata, water levels, aquifer type, and yields, etc.

Based on the existing and additional data generated, various spatial maps on water level, quality, and subsurface lithological sections, which included 2-D, 3-D, and fence diagrams, were prepared. This was followed by the preparation of aquifer maps on a 1:50,000 scale and the formulation of implementable management plans [2].

3. Case study area

Nashik district is one of the six districts of the *Khandesh region* of Maharashtra state (Latitude: 19°35'N and 20°50' N; Longitude: 73°16′ E and 74°56′ E). Nashik district is a typical volcanic basaltic area of Maharashtra state, India. Maharashtra state occurs in the western part of India, whereas Nashik district is located in the northwestern part of Maharashtra state (**Figure 1**). It is the third largest district in Maharashtra in terms of population of 61,09,052 and occupies an area of 15,582 sq km in the north-west part of Maharashtra. The Nashik district, one of the largest districts of the state, has 15 towns/blocks and 1930 villages. Nashik is surrounded by almost 08 districts *viz*. Dhule, Jalgaon, Aurangabad, Ahmednagar, Thane, and Valsad etc. of Maharashtra and the Navsari and Dangs districts of Gujarat. Thus, eight districts adjoining Nashik form the boundary between Maharashtra and Gujarat. The major part of the district comes under Godavari and Tapi basins. Godavari originates from the Brahmagiri mountain range in the *Western Ghats* of Nashik district of Maharashtra state [4].

Climate and rainfall play important roles in replenishing groundwater resources, especially the shallow aquifer in such semiarid areas, where rainfall is the primary and main source of recharge. Nashik district experiences a hot tropical climate with extreme summer, mild winter season, and general dryness throughout the year except during the southwest monsoon season, that is. June to September. The rainfall in the district is under the influence of the southwest monsoon. Nashik district falls 'assured rainfall zone' under agro climatic zones. The maximum temperature in summer is 42.5°C and the minimum temperature in winter is less than 5.0°C. Relative humidity ranges from 43 to 62%.

The normal annual rainfall in the district is uneven and varies from 550 mm to 3400 mm (**Figure 2**). The average rainfall of the district is ≈ 1043.6 mm, spread over 69 rainy days. Igatpuri receives the highest annual rainfall in the district. The presence of both the Western Ghat (hills) and the plateau region, together with the monsoon winds and rugged terrain condition forms the *rain shadow zones* that result in scanty and deficient rainfall.

In the study area, the major part is covered by Deccan Trap basalt aquifers comprising two aquifer systems, whereas alluvium has restricted to only 1500 sq. km. area

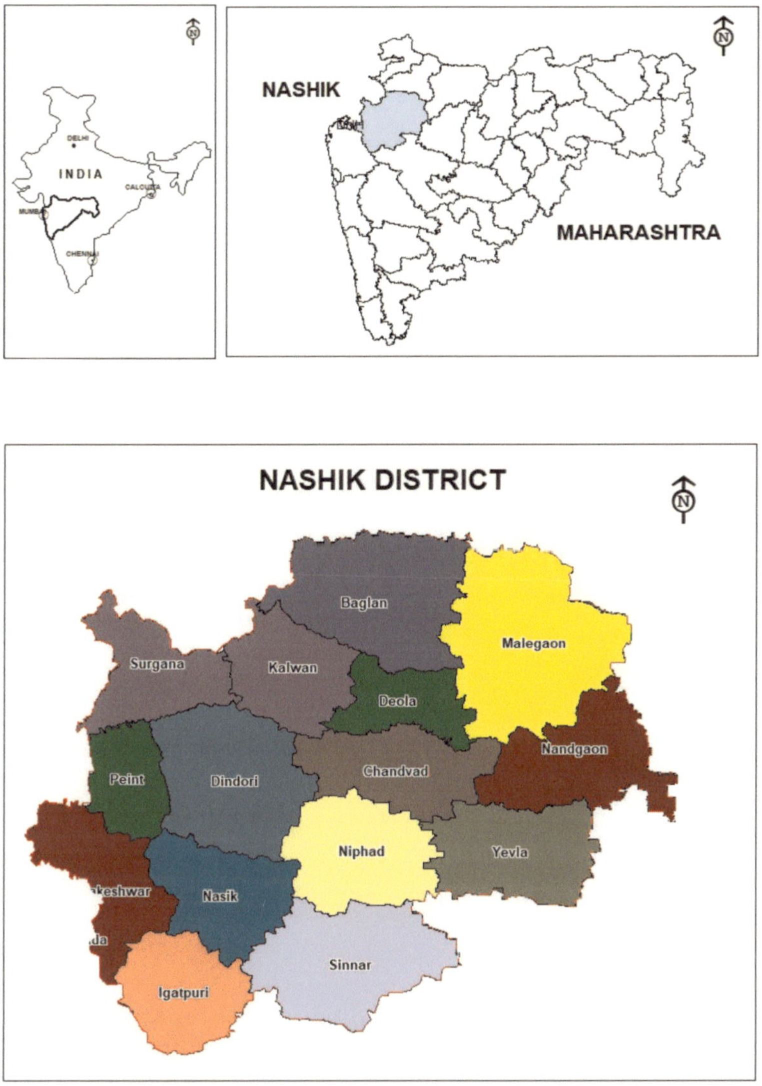

Figure 1.
Index map.

along the banks of major rivers and considering its limited thickness only one aquifer is established in alluvium.

4. Aquifer system

The entire area of the district is underlain by the basaltic lava flows of the upper Cretaceous to lower Eocene age forming a hard rock aquifer. The shallow alluvial formation of the recent age also occurs as a narrow stretch along the banks of *Godavari* and *Girna* Rivers flowing in the area forming a soft rock aquifer.

The groundwater data generated from the monitoring wells, micro-level hydrogeological inventories, exploratory and observation wells, and various thematic

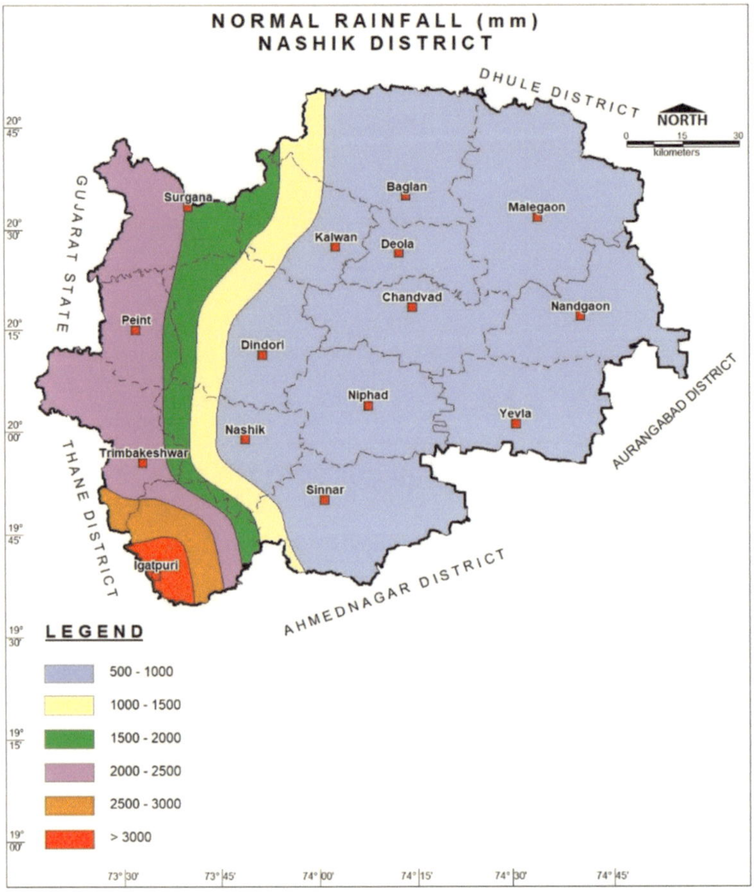

Figure 2.
Rainfall distribution.

layers were utilized to decipher the aquifer disposition of the area. This substantiates information on the aquifer geometry and hydrogeological information about the region where these aquifers occur. In the area, alluvium and Deccan Trap basalt are the only formations. In Deccan Trap basalt, two aquifer systems have been deciphered, whereas in alluvium considering its limited thickness only one aquifer is established (**Figure 3**) as listed below:

I. **Soft Rock Alluvium**

Aquifer – I (Shallow Aquifer): 8 to 32 m

II. **Hard Rock Deccan Trap Basalt**

a. Aquifer – I (Shallow Aquifer): 8 to 32 m

b. Aquifer – II(Deeper Aquifer): 30 to 200 m

The aquifer system characteristics are described in detail in this chapter as the management options mainly depend on these characteristics [1, 2].

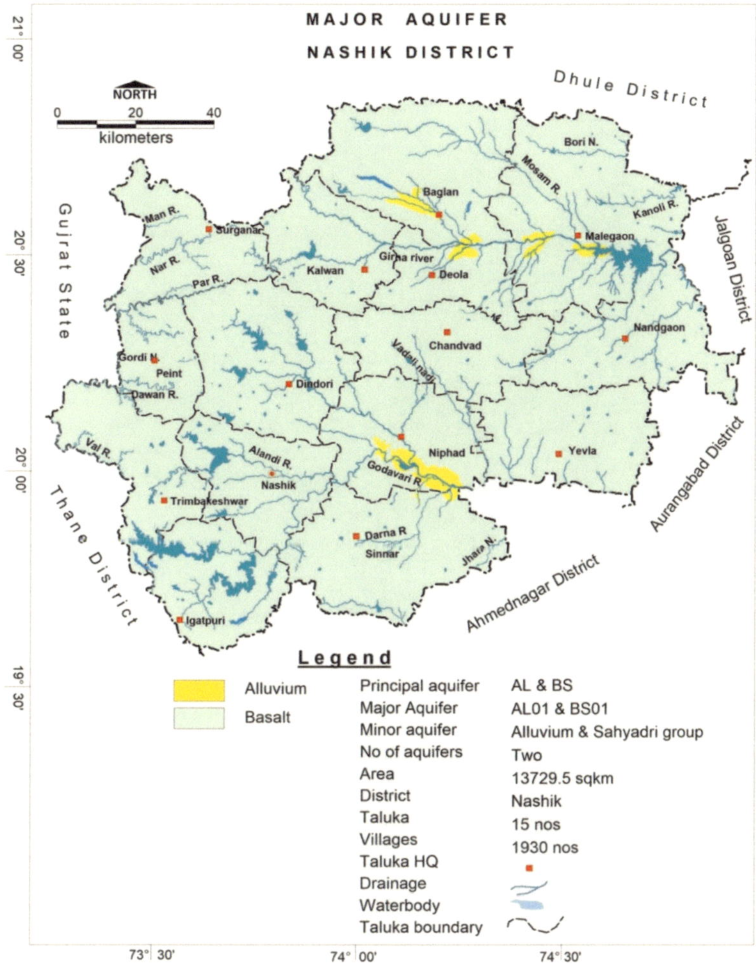

Figure 3.
Major aquifers.

4.1 Soft rock aquifer - alluvium

Alluvium occurs in small areas in the form of discontinuous patches along the banks and flood plains of major rivers such as Godavari, Girna, and their tributaries. In alluvium, the granular detrital material, such as sand and gravel, usually occurs as a thin layer in the district that yields water. In the district, alluvium occupies an area of 1500 sq. km consisting of reddish and brownish clays with intercalations of sand, gravel, and Kanker. The loosely cemented coarse sands and gravels form 3–4 meters thick lower most horizons at the bottom of these alluvial pockets. Groundwater in alluvium occurs under unconfined conditions. The dug wells constructed in alluvium are ranging in depth from 8 to 15 m, whereas the borewells range in depth from 15 to 25 m and the yield of both the dug wells and bore wells ranges from 13 to 22 m^3/day.

In the alluvium aquifer, the narrow deposits along river courses are observed in the northern part of 15 to 20 m thickness, whereas in the southern part near Godavari

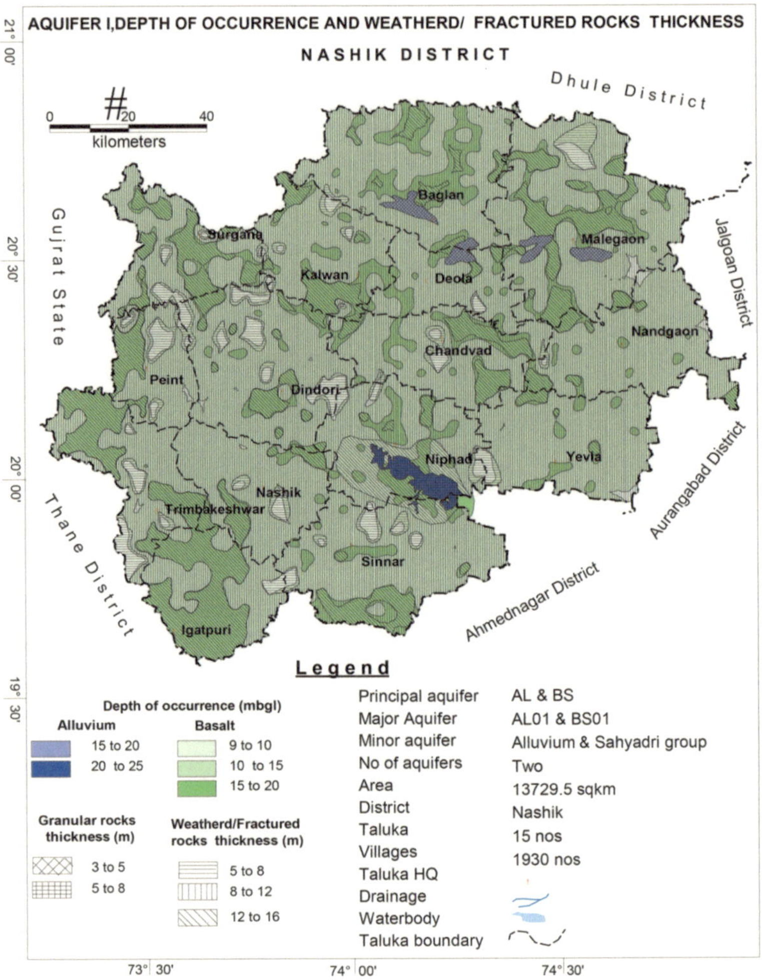

Figure 4.
Thickness and depth of occurrence of aquifer – I.

River, it is observed in 20 to 25 m depth range with a thickness of granular material being 5 to 8 m (**Figure 4**).

4.2 Hard rock aquifer – deccan trap basalt

Basaltic lava flows are normally horizontally disposed over a wide stretch and give rise to a plateau. These flows occur in layered sequences and are represented by a massive unit at the bottom and a vesicular unit at the top of the flow. Flows are separated from each other by a marker bed known as a 'bole bed,' which is formed due to weathering of the top part of the flow and represents a hiatus in volcanic activity. The bole beds indicate the change in basaltic flows.

The groundwater in the Deccan Trap occurs mostly in the upper weathered and fractured parts down to 8 to 32 m depth. At places, potential zones are encountered at deeper levels in the form of fractures and inter-flow zones. The upper weathered and fractured parts form a phreatic aquifer and groundwater occurs under water table (unconfined) conditions. The water levels range from 1.2 to 15 m bgl and yield varies from 10 to 100 m^3/day depending upon the local hydrogeological conditions. At deeper levels, the groundwater occurs under semi-confined to confined conditions. Borewells drilled down to 200 m depth, tapping weathered, vesicular, and fractured basalt yielded negligible to 2.5 liters per second (lps), whereas the water levels varied from 8 to 55 m bgl.

Deccan basalts are hydrogeologically heterogeneous rocks. The weathered, jointed, and fractured parts of the rock constitute the zone of groundwater storage and flow. The existence of multiple aquifers is characteristic of basalt and is indicative of wide variation in the joint/fracture pattern and intensity. The yield of wells is a function of the permeability and transmissivity of the aquifer, and it depends upon the degree of weathering, the intensity of joints and fractures, and the topographic setting of the aquifer. Due to wide variations in secondary openings, the potential areas for groundwater are generally localized. In general, groundwater occurs under phreatic/ unconfined to semi-confined conditions in basalts.

The perusal of the Aquifer-I map (**Figure 4**) indicates the major part of the shallow basaltic aquifer is observed in the 10 to 15 m range, and the thickness of the aquifer is 8 to 12 m range. The depth of occurrence of 15 to 20 m with 12 to 16 m weathered and jointed thickness in the basaltic aquifer is observed in discontinued patches in the entire district; however, in the western fringe part of the district it is observed prominently and it coincides with highly dissected plateau and denudational slopes. The less than 10 m depth of occurrence is observed in limited and small areas. The Aquifer- I map indicates that a moderate depth of aquifer is available in the major part of the district with sufficient aquifer thickness.

The Aquifer – II is observed only in Deccan Trap basalt formation. The map showing spatial disposition and the vertical extent of Aquifer-II indicating its depth of occurrence and fractured/granular rock thickness has been generated and shown in **Figure 5**. The Aquifer-II map indicates the deeper basaltic aquifer in the major part (100 m to 140 m range and thickness 1 m to 3 m). The depth of occurrence of 140 to 178 m is observed in the eastern part of the district in Nandgaon and Yeolablocks, in the small central part with 9 to 12 m fractured thickness. The south-western and north-eastern parts of the district are having a shallower depth of occurrence of 30 to 80 m and fractured thickness of 0.50 to 1.00 m in south-western parts and fractured thickness of less than 0.50 m in north-eastern parts of the district. The Aquifer II map indicates that the deeper aquifer depth of occurrence in the major part of the district is 80 to 140 m bgl and the fracture thickness of 1 to 3 m is most vastly spread in the district.

4.3 Yield potential of aquifer-I and II

The yield potential of the aquifer is the capacity of the aquifer to yield groundwater. The yield potential of Aquifer-I is plotted in **Figure 6** and it indicates that in the entire basaltic terrain, the yield is less than 15 m^3/day. However, if we correlate the yield potential of Aquifer-I with the depth of occurrence and weathered thickness, it indicates that even though the aquifer occurs down to a moderate depth of 10 to 15 m with a sufficient thickness of 8 to 12 m, and it does not guarantee adequate yield in hard rock basaltic aquifer. Thus, it can be concluded that the yield potential of the basaltic Aquifer - I is independent of aquifer thickness.

The yield potential of Aquifer-II is plotted in **Figure 7** and it proves that the major area is characterized by low yields approximately ranging between 0.5 to 1 lps,

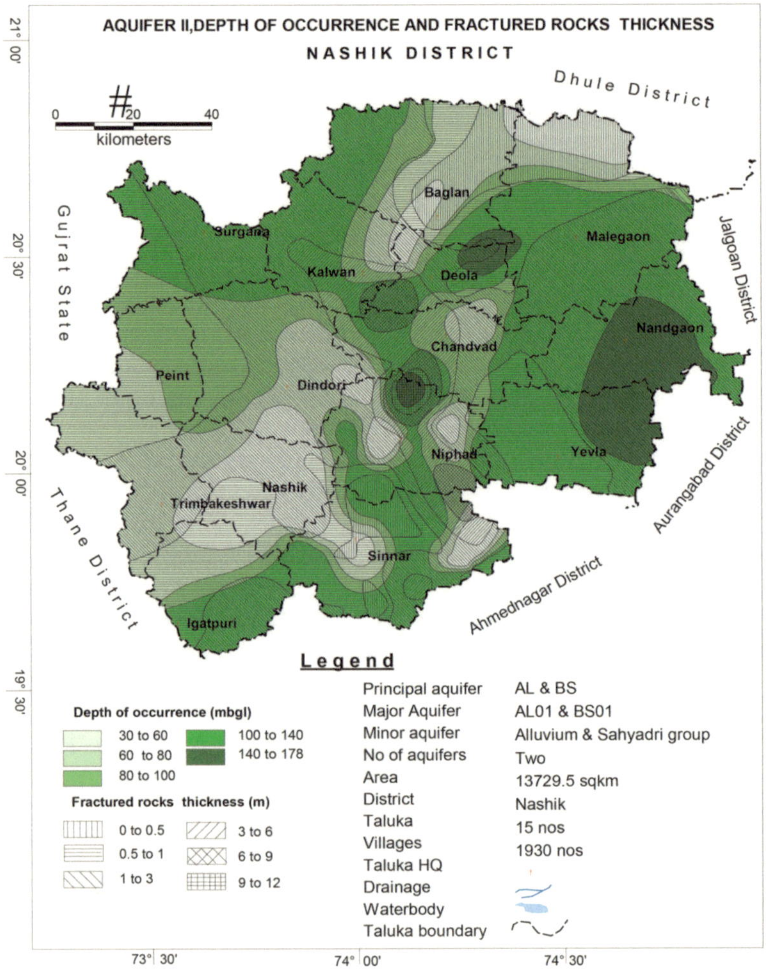

Figure 5.
Thickness and depth of occurrence of aquifer – II.

whereas moderate yields of 1.00 to 1.50 lps are observed in around 25% of the area. In a very small part of Niphad taluka near Godavari River moderate to high yield of 1.50 to 2.50 lps is recorded. However, if we correlate the yield potential of Aquifer-II with a depth of occurrence and fractured thickness, it indicates that even though the aquifer occurs down to moderate to deep depths of 80 to 140 m and is present in a major part of the area, but the yield is not dependent on the depth of occurrence of the aquifer. Thus, the common tendency of the borewell culture of going deep for getting more groundwater is not true in hard rock areas, especially basaltic aquifers.

4.4 3-D aquifer disposition

Based on the existing data, Mapinfo software 8.5 was used to prepare aquifer disposition in 3D, fence diagram, and 3D Lithological disposition diagram, and several hydrogeological sections have been prepared along section lines to understand the subsurface disposition of the aquifer system.

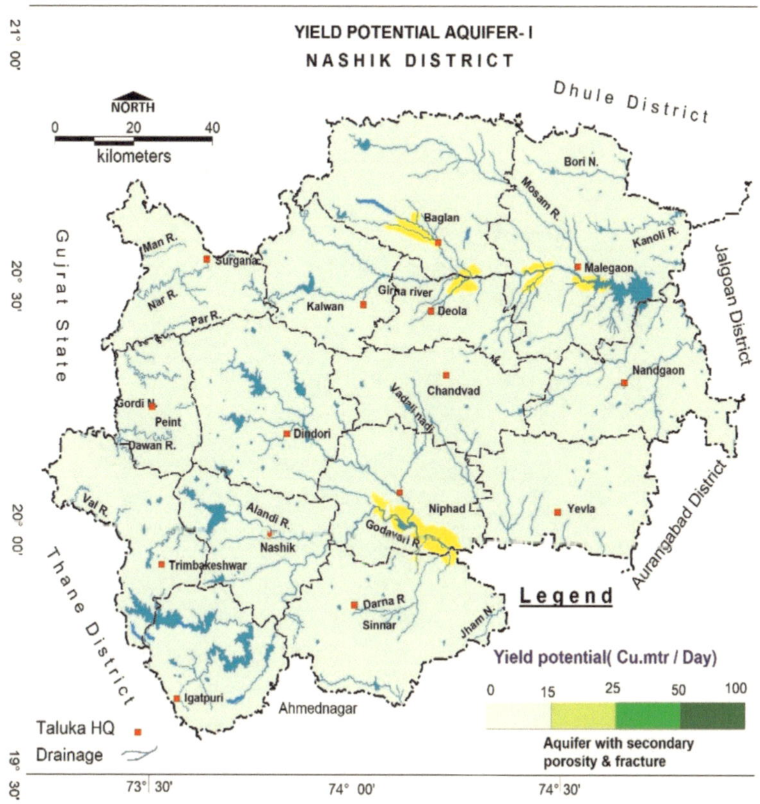

Figure 6.
Yield potential of aquifer – I.

The 3D aquifer disposition is presented in **Figure 8**, which helps in visualizing the disposition of the aquifers in the three-dimensional model with topographic elevation and drainage. The 3D aquifer fence diagram and the aquifer bar diagram are shown in **Figures 9** and **10**. Both the fence and bar diagram covers the entire district and they reassert the factual position of the aquifer -I being shallow in depth with less thickness as compared to the aquifer -II. Both these figures also show that the massive part is the most dominant formation in vertical disposition limiting the occurrence of the water-bearing aquifers above it. The fence diagram gives a clear picture regarding the prospective depth of drilling the borewell at any given location in the district.

4.5 2-D hydrogeological cross section

To study the aquifer disposition in detail, four hydrogeological cross sections indicating aquifer geometry in different directions have been prepared, and the section lines are plotted in **Figure 9**. These sections indicate the disposition of Aquifers – I and II along with fracture and water levels.

4.5.1 Hydrogeological cross section A-A'

Hydrogeological cross section A-A' (**Figure 11**) represents data from seven exploratory wells for 136 km in the N-S direction with the depth of cross section

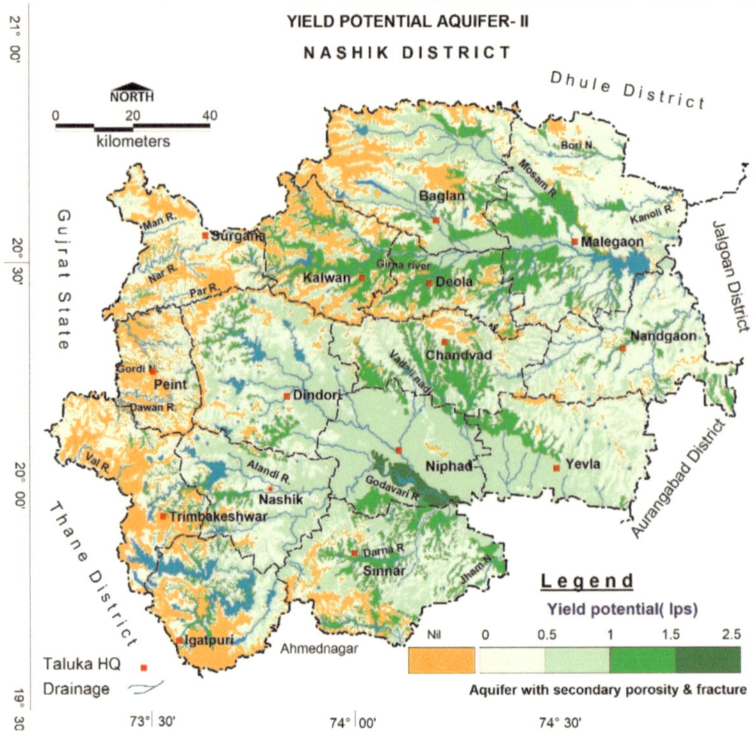

Figure 7.
Yield potential of aquifer – II.

varying from 90 m bgl to 202 m bgl. The thickness of the Aquifer-I is almost uniform along the section line, whereas the Aquifer–II is having more thickness near Washala in the northern part and Kankapur in the central part of the section with the discharge/yield of 8.77 lps at Kankapur. Southwards from Kankapur along the section line, it can be seen that the thickness of Aquifer-II (deeper aquifer) is decreasing. The

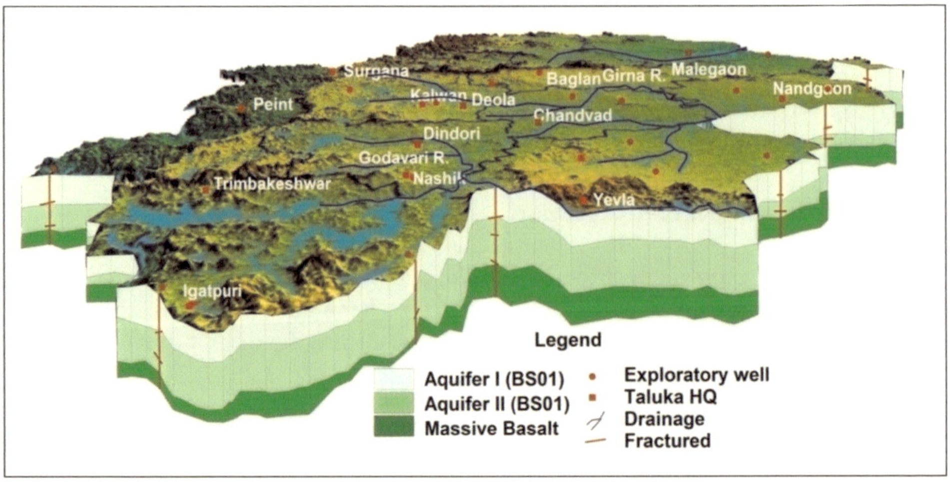

Figure 8.
3D aquifer disposition.

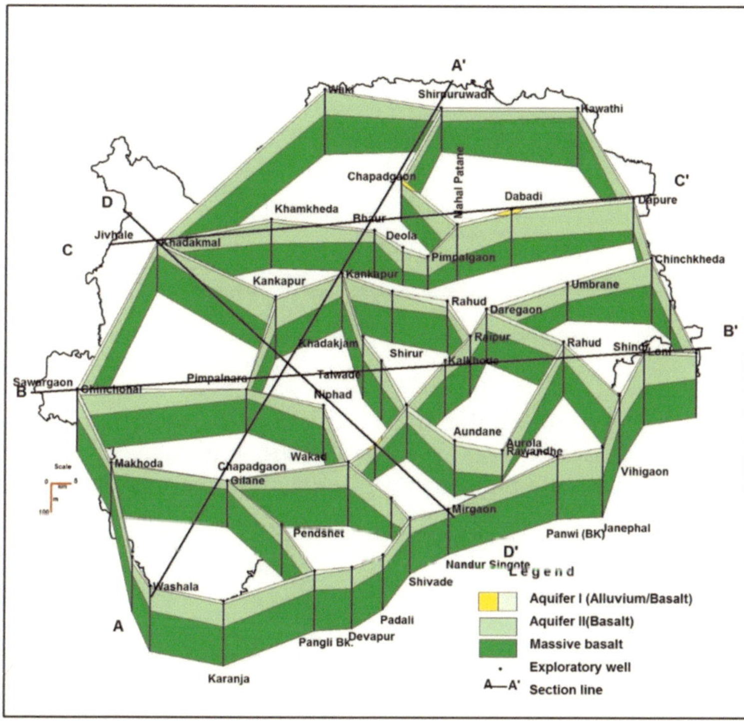

Figure 9.
Fence diagram.

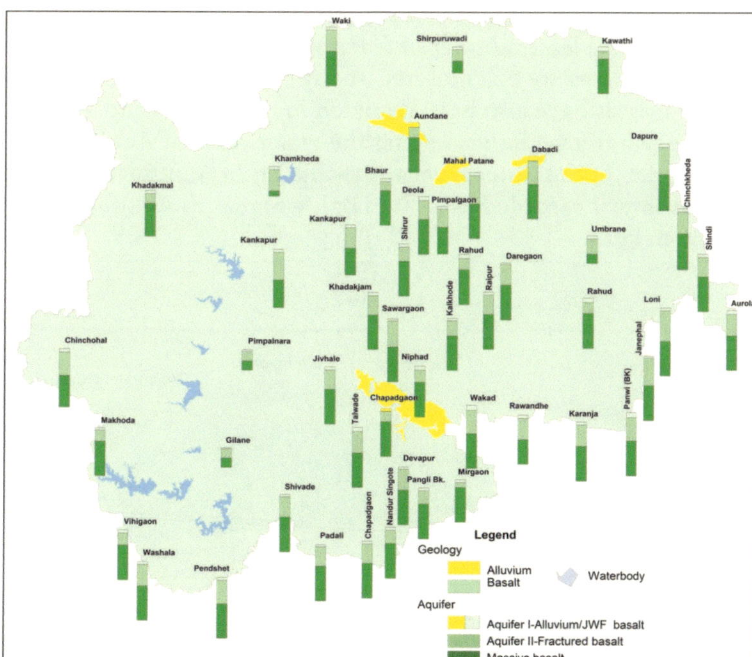

Figure 10.
Point aquifer disposition by Bar diagram.

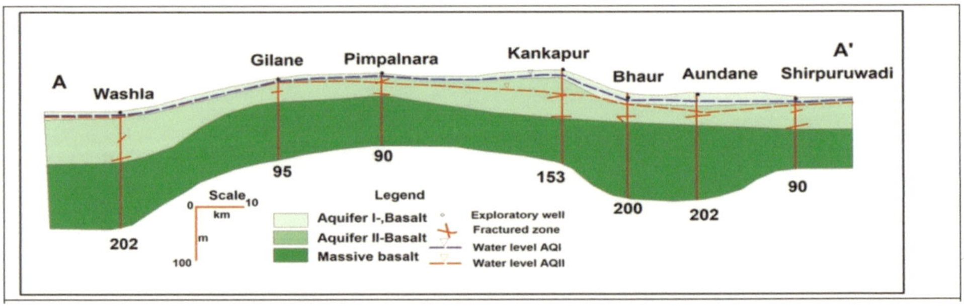

Figure 11.
Lithological section AA'.

water levels of Aquifer-I and Aquifer-II have also been depicted in the section and a close observation of the water level indicates that the water table of Aquifer-II is almost the same at Gilane and Shirpurwadi, whereas at other places it is below the water level of Aquifer-I.

4.5.2 Hydrogeological cross section B-B′

Hydrogeological cross section B-B′ (**Figure 12**) represents data from six exploratory wells with a depth of cross section ranging from 180 m bgl to 200 m bgl for 163 km in the W-E direction in the southern part of the district. The thickness of the Aquifer-I is almost uniform along the section line, whereas Aquifer II is having more thickness in the western part up to Sawargaon after which it starts decreasing till Rahudin central-eastern parts. The maximum thickness of Aquifer – II and the maximum number of fractures encountered are observed at Loni in the central part of the section. At Kalkhode, the least number of fractures (1) is encountered, whereas at all other locations, 2 fractures are encountered at various depths. The water levels of Aquifer-I and Aquifer-II have also been depicted in the section and a close observation of the water level indicates that the water table of Aquifer-II is below the water level of Aquifer-I. The water level at Sawargaon in Aquifer-I during the pre-monsoon season was recorded as 7.60 m bgl, whereas in Aquifer-II it was recorded as 13.56 m bgl.

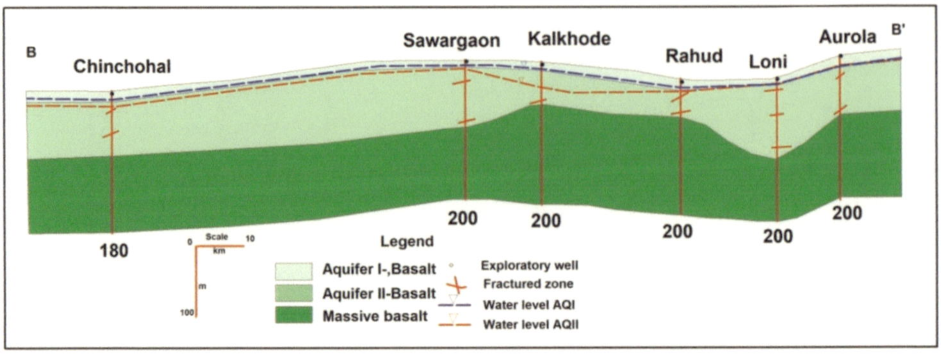

Figure 12.
Lithological section BB'.

4.5.3 Hydrogeological cross section C-C′

Hydrogeological cross section C-C′ (**Figure 13**) represents data from six exploratory wells with the depth of cross section varying from 190 m bgl to 203 m bgl for 126 km WE direction in the northern part. The thickness of the Aquifer-I and Aquifer–II is less in the western part from Khadakmal to Bhaur, whereas in the eastern part of the section line, both the aquifers are having more thickness from Bhaur to Dapure. At Dapure, a maximum number of fractures are encountered with the discharge/yield of 3.17 lps and at Bhaur, the least number of fractures (1) is encountered with a discharge of 0.14 lps, whereas at all other locations, two fractures are encountered at various depths.

4.5.4 Hydrogeological cross section D-D′

Hydrogeological cross section D-D′ (**Figure 14**) represents data from five exploratory wells with the depth of cross section varying from 156 m bgl to 200 m bgl for 106 km in the NW – SE direction. The thickness of the Aquifer-I is more in the southwestern part and southeastern parts of the section from Khadakmal to Jivhale and from Chapadgaon to Mirgaon. The thickness of Aquifer – II is more in the NW part of the section line from Khadakmal to Jivhale and decreases in the southeastern part of the section near Mirgaon, where discharge/yield of 1.37 lps has been encountered. The massive thicker part in this section indicates the limited thickness of the deeper aquifer.

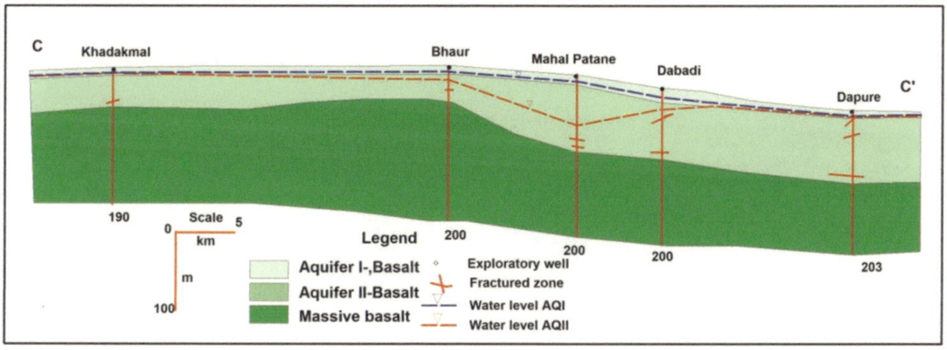

Figure 13.
Lithological section CC'.

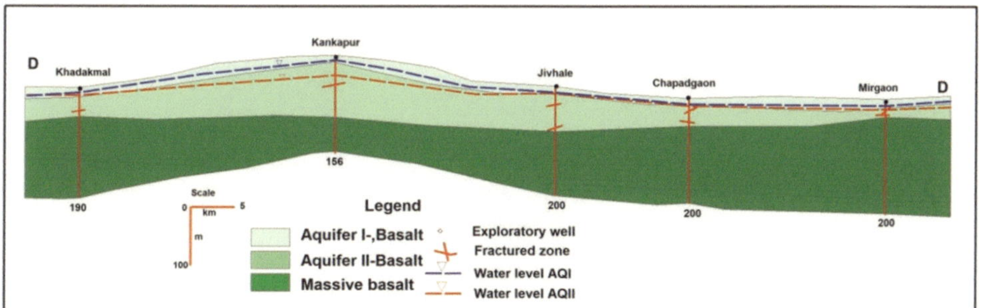

Figure 14.
Lithological section DD'.

4.6 Pre-monsoon depth to water level (may 2021) of shallow aquifer

The depth to water levels in Nashik district during May 2021 ranges between 0.05 (Trambakeshwar, Trambakeshwar block) and 30.00 m bgl (RavalgaonPz, Malegaon block). The depth to water levels of more than 10 m bgl is observed in mainly Baglan and Kalwan blocks and isolated patches in the remaining parts of the district covering an area of 2118.53 sq. km. The depth to water level between 2 and 5 and 5–10 m bgl is recorded in the major areas of the district covering an area of 12799.4 sq. km. Water level ranges between 0 and 2 m bgl are observed in isolated patches in the Nashik

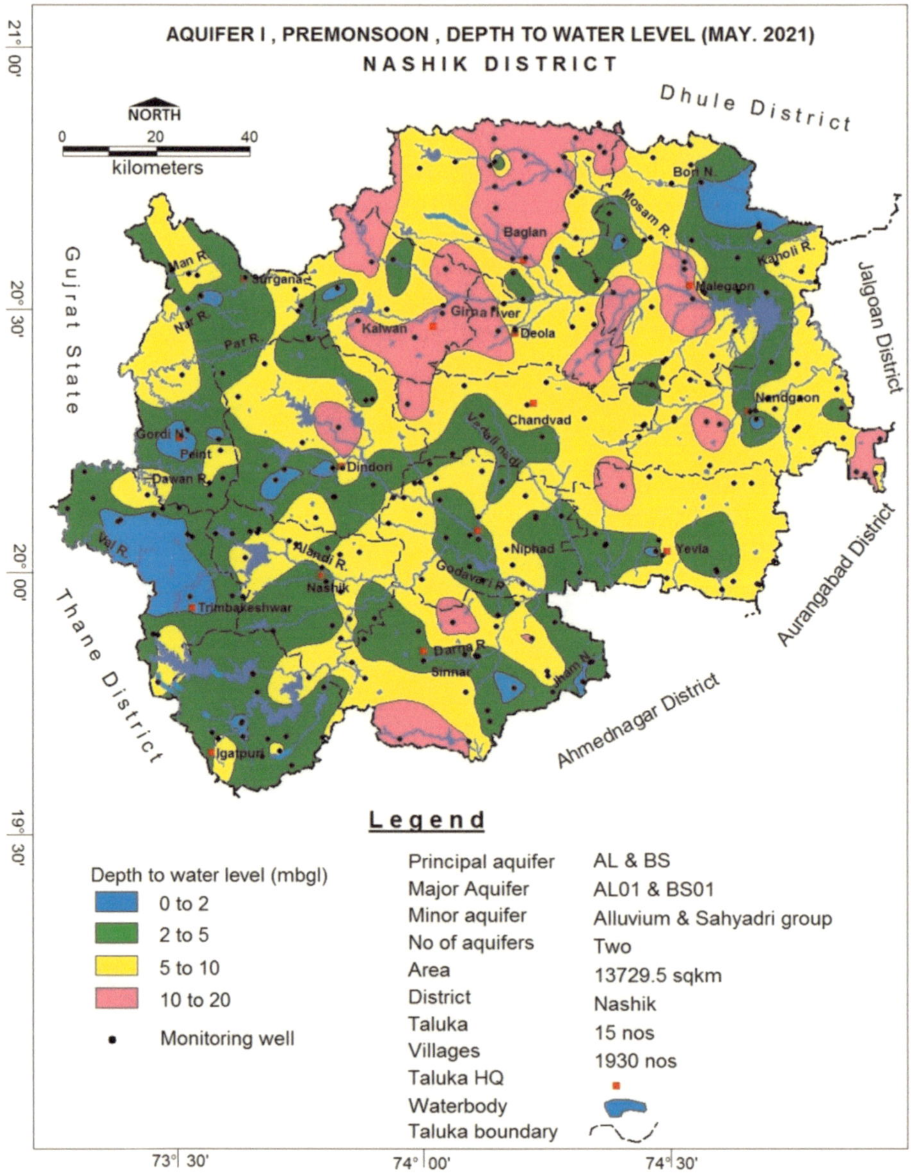

Figure 15.
Pre-monsoon (may 2021) depth to water level of shallow aquifer.

district covering an area of 675.31 sq. km. The pre-monsoon depth to water level map is depicted in **Figure 15**.

4.7 Post-monsoon depth to water level (Nov 2021) of shallow aquifer

The depth to water levels in Nashik district during November 2021 ranges between ground level (PathareBk, Sinnar block, Chindhi, Dapur, and Nagzari in Malegaon block) and 17.00 m bgl (Chandanpuri, Malegaon block). The depth to water level up

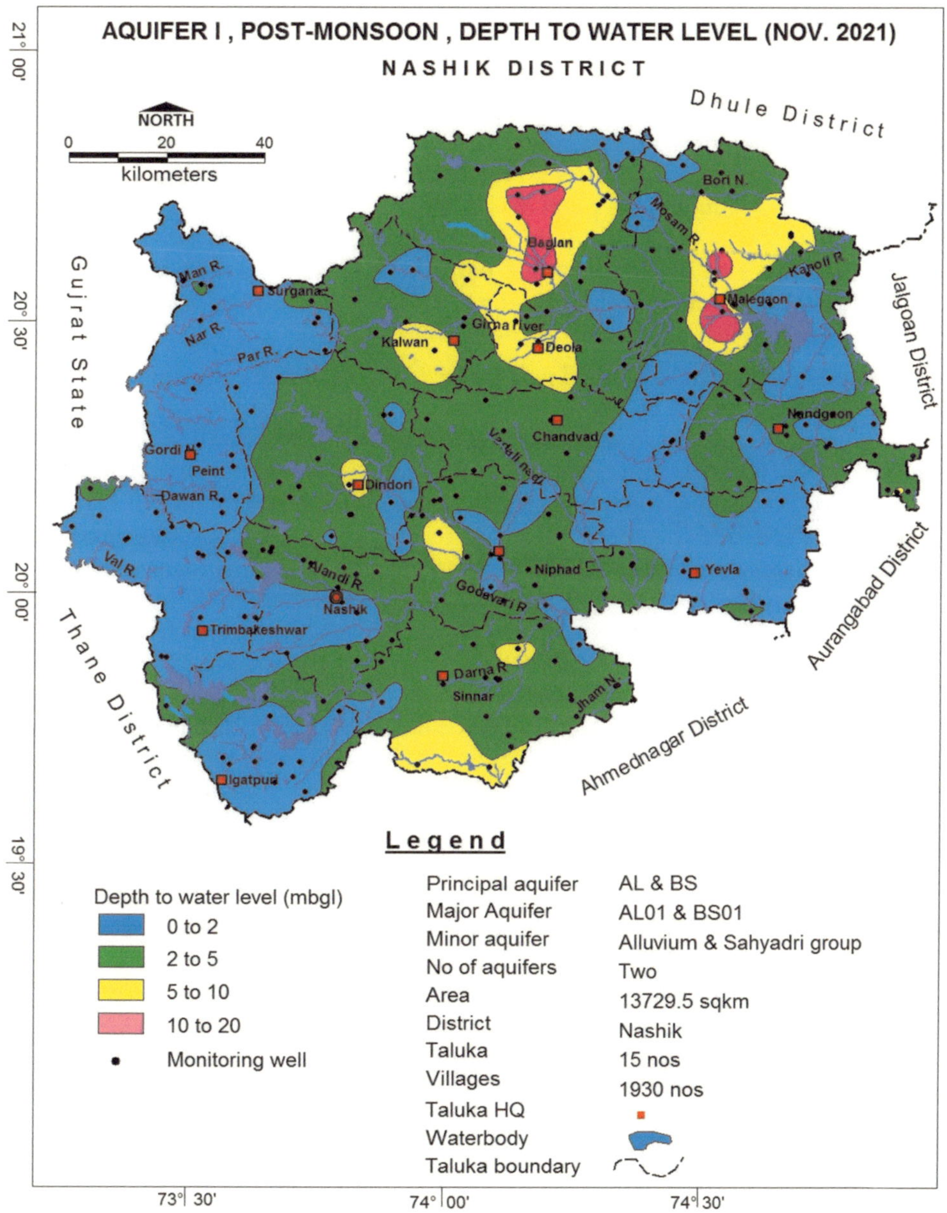

Figure 16.
Post-monsoon (Nov. 2021) depth to water level of shallow aquifer.

to 2 m and 2 to 5 m bgl covers a major part of the district covering an area of around 13890.32 Sq.km. Water level range between 5 to 10 m bgl is observed mainly in Baglan, Deola, and Malegaon blocks and isolated patches in Kalwan, Niphad, Dindori, Sinnar, and Chandwad blocks covering an area of 1487.99 Sq.km.The depth to water levels of more than 10 m bgl are observed in Baglan and Malegaon blocks in the district covering an area of 214.67 Sq.km. The post-monsoon depth to water level map is depicted in **Figure 16**.

4.8 Water level variation between pre-monsoon (may 2021) and post-monsoon (Nov.2021)

The water level variation between the pre-monsoon and the post-monsoon season was also analyzed to study the consequence of the rainfall on the water level scenario. It is noticed that the entire area shows a rise in water levels in the post-monsoon season due to recharge from monsoon rainfall. The minimal water level variation of 0.10 m was observed in various parts of the Baglan, Nandgaon, Niphad, and Trimbakeshwar blocks, while the maximal water level variation of 12.35 m was measured at Visapur, Yevla block. The major part of the area shows the water level variation from 2 to 5 m followed by 5–10 m.

4.9 Pre-monsoon depth to water level (may 2021) of deeper aquifer

Pre-monsoon depth to the water level in Nashik district during May 2021 ranges from 8.76 m bgl (Hingalwadibk, Kalwan block) to 123.5 m bgl (Kaluste, Igatpuri block). The pre-monsoon depth to water level map for Aquifer -II presented in **Figure 17** indicates that the depth to the water level of less than 10 m bgl is observed in mainly Surgana, Peth and small isolated patches in Klawan and Dindori blocks covering an area of 626 Sq.km. The major parts of Peth, Dindori, Kalwan, and Baglan blocks and small isolated patches in Yevla, Niphad, Nashik, and Trambakeshwar show depth to water level between 10 and 20 m bgl covering an area of 2855.37 Sq.km. The deeper water level of more than 50 m bgl is observed in major parts of the district in Igatpuri, Sinnar, Nashik, Trambakeshwar, Niphad, Nandgaon, Malegaon, Deola, and Chandwad blocks covering an area of 7097.14 Sq.km. This may be due to the overexploitation of groundwater and it also indicates that the shallow aquifer is not able to sustain the demand; hence, groundwater resources are being also withdrawn from deeper aquifers in this area.

4.10 Post-monsoon depth to water level (Nov 2021) of deeper aquifer

In Aquifer-II, the post-monsoon depth to water level in Nashik district during Nov. 2021 ranges between 3.9 m bgl (Sinnar, Sinnar block) and 71.6 m bgl (Sakore, Nandgaon block). The post-monsoon depth to water level for Aquifer -II is given in **Figure 18** and it shows that the depth to water level of less than 10 m bgl has been demarcated in Peth, Surgana, Nashik, Dindori, and parts of Niphad, Sinnar, Kalwan, and Trambakeshwar blocks covering an area of 3647.15 Sq.km. Depth to water level between 10 m bgl to 20 m bgl has been observed in the major part of the district covering an area of 4381.61 Sq.km. Southern parts of the district near Sinnarand eastern parts of the district near Nandgaonshow the deepest water level of more than 50 m bgl covering an area of 2348.13 Sq.km. Deeper water levels between 20 and 50 m

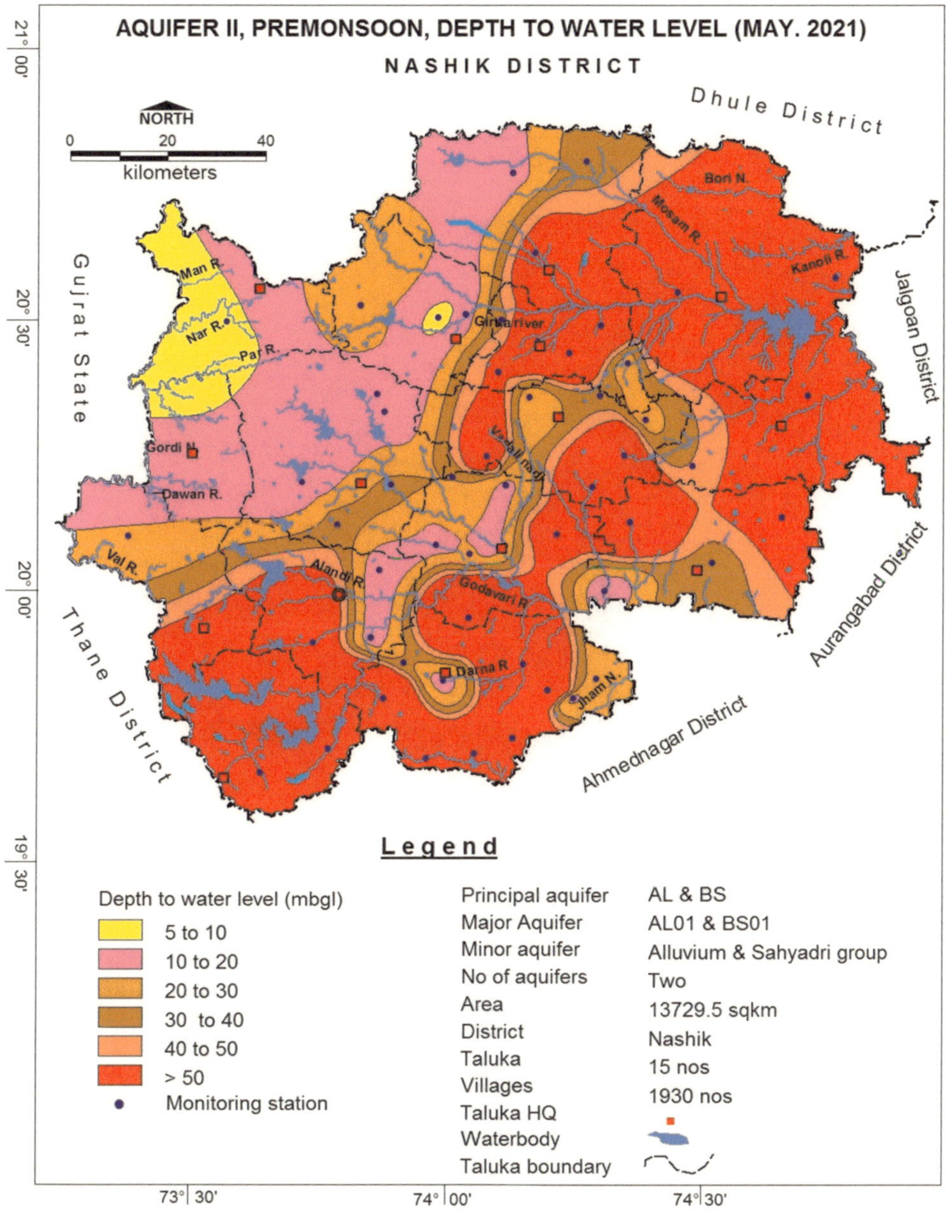

Figure 17.
Pre-monsoon depth to water level of deeper aquifer.

bgl are observed in the central-eastern parts of the district near Baglan and Chandvad and also in the south-western parts of the districts.

4.11 Water level trend (2012-2021)

The behavior of water level with time gives information on whether water levels are rising or falling with time and the rate of rise and fall. The water level trend is

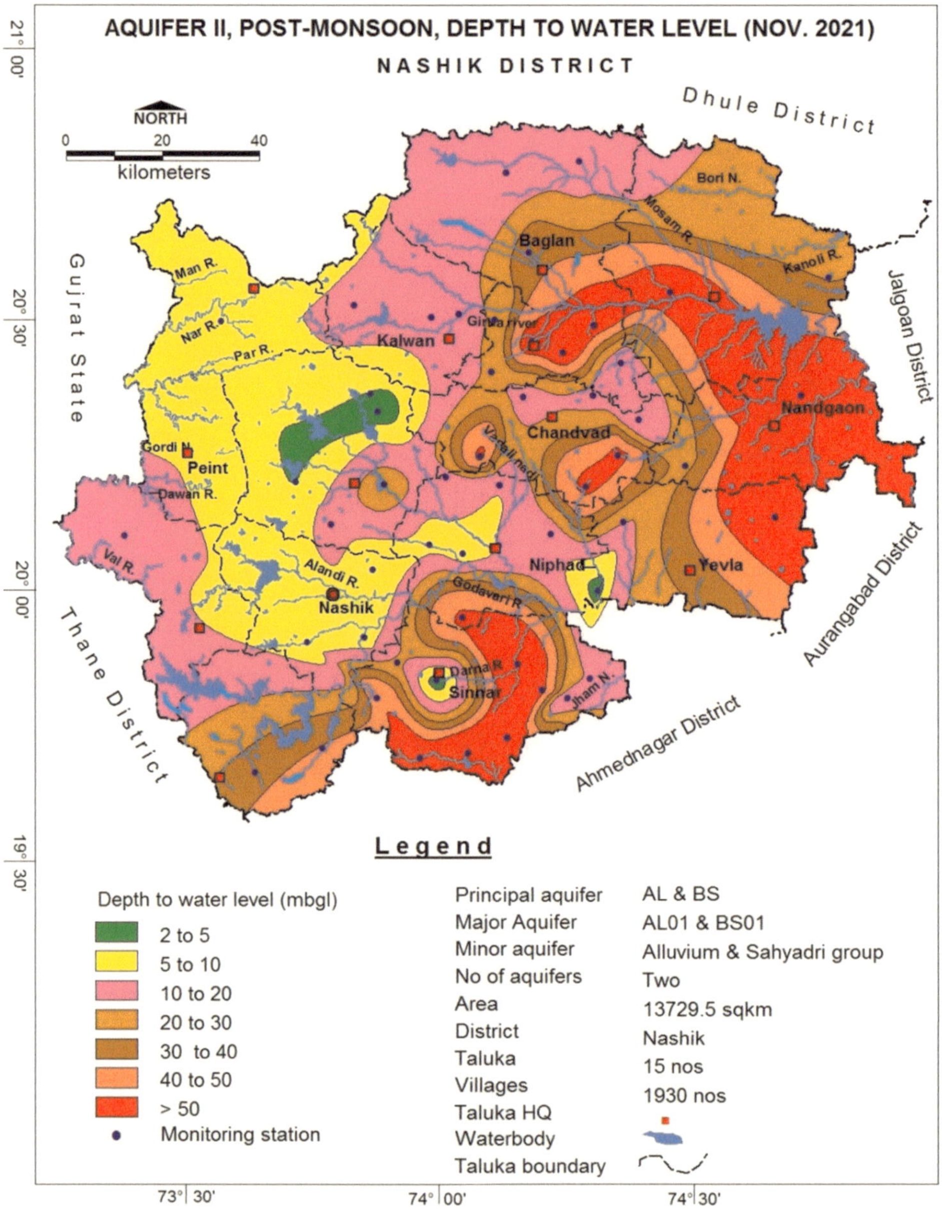

Figure 18.
Post-monsoon depth to water level of deeper aquifer.

obtained by time series analysis by plotting time on the X-axis and water level on the Y-axis, the trend line is added to the resultant plot, which gives the water level trend for that particular well. Time series analysis helps understand the underlying causes of trends or systemic patterns over time.

During pre-monsoon, a rise in water level trend has been recorded at 105 monitoring wells and it ranges from 0.003 m/year (Padalde, Malegaon block) to 0.99 m/year (Sakore, Kalwan block), while the falling trend was observed in 80 monitoring wells varying from −0.003 (Hatrundi, Surgana block) to −1.49 m/year

(Jaidar, Kalwan block). The rising trend of more than 0.2 m/year is observed in parts of the Kalwan, Niphad, and Nasik blocks covering an area of 3456.13 km^2. The falling trend of more than 0.2 m/year is observed in northern peripheral areas especially east of Surgana and western peripheral areas near Peint covering an area of 2832.01 km^2 (**Figure 19**). During post-monsoon, a rise in water level trend has been recorded at 145 stations and it ranges between 0.0006 m/year (PimpalgaonMor, Igatpuri block) and 1.01 m/year (Chirai, Baglan block), while the falling trend was observed in

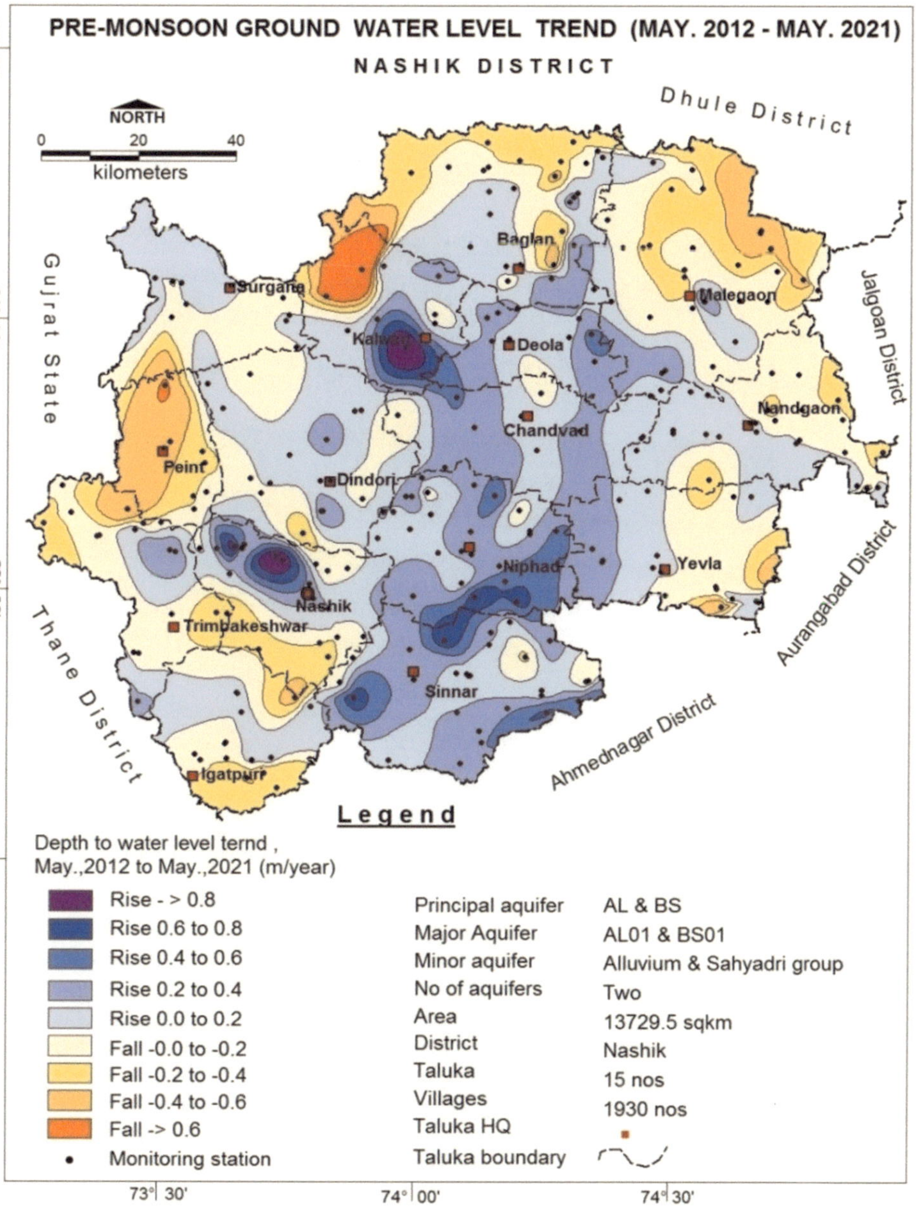

Figure 19.
Pre-monsoon decadal water level trend (2012-2021).

36 stations varying from 0.0006 (Ankai, Yevla block) to 0.33 m/year (Sakur, Igatpuri block). The rising trend of more than 0.2 m/year is observed near Deola, Kalwan, Chandwad, and Niphad covering an area of 4938.42 km^2. Whereas a falling trend of more than 0.2 m/year is observed in a limited area of 216.39 km^2 in the eastern part of Malegaon and the northern part of Igatpuri (**Figure 20**). The rising water level trend indicates that the groundwater recharge during the period is adequate considering the storage capacity of the aquifer and is more than withdrawal, whereas falling trend indicates that the groundwater withdrawal is more than the recharge.

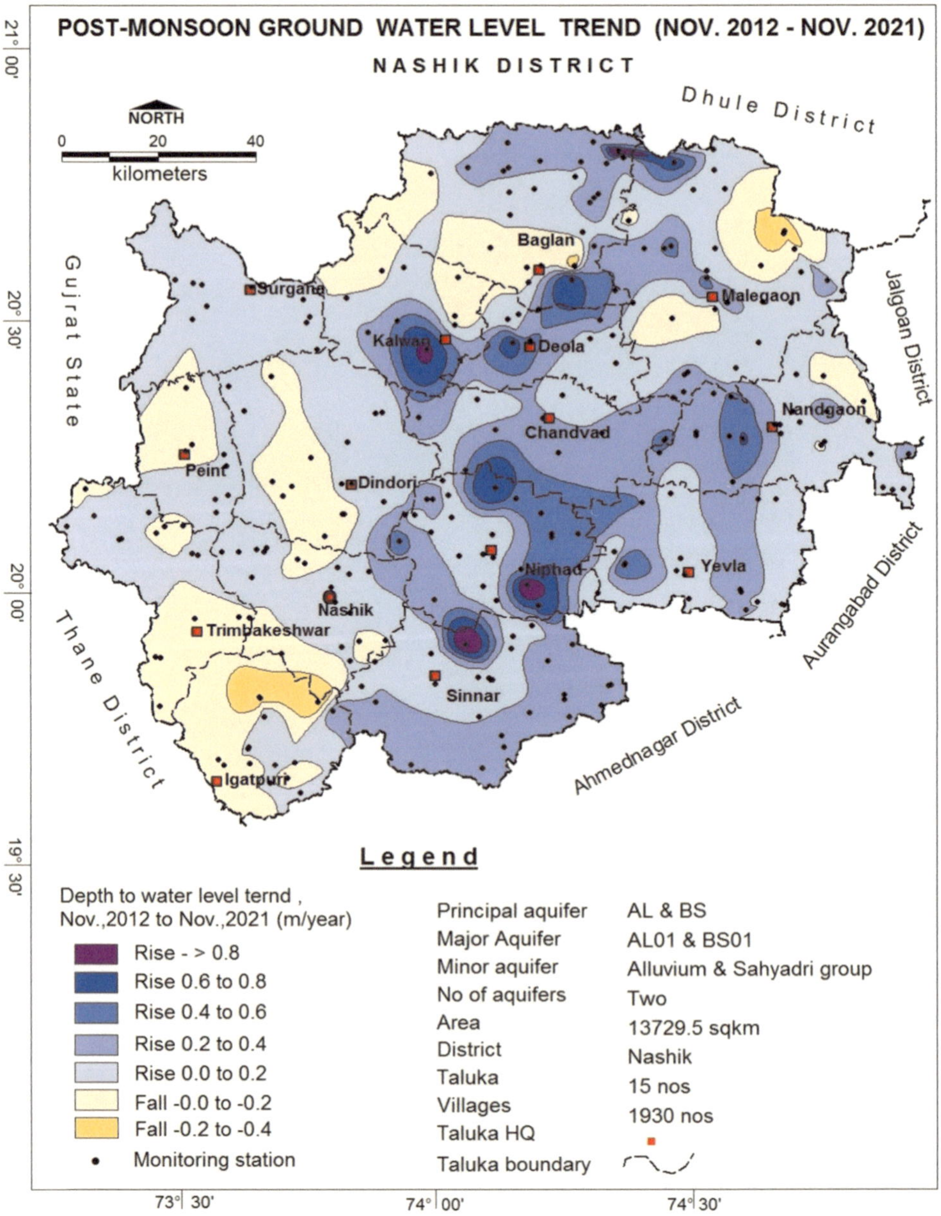

Figure 20.
Post-monsoon decadal water level trend (2012-2021).

Type of Aquifer	Aquifer-I (Hard rock-Deccan Trap Basalt)	Aquifer-I (Soft Rock - Alluvium)	Aquifer-II (Hard rock-Deccan Trap Basalt)
Formation	Weathered/Fractured Basalt	Sand Clay Gravel Kankers etc.	Jointed / Fractured Basalt
Depth of Occurrence (m bgl)	8 to 32	8 to 25	30 to 200
Granular/Weathered / Fractured rocks thickness (m	1 to 14	7 to 21	0.5 to 12
SWL (mbgl)	1.40–21.00	5.0 to 10.0	8 to 55
Yield	10 to 100m^3/day	200–300 m^3/day	Up to 3.17 lps
Sustainability	1.0 to 3 hrs	2 to 5 hrs	0.5 to 3 hrs
Transmissivity (m^2/day)	9.25 to 89.04 m^2/day		10.85 to 131.11 m^2/day
Specific Yield/Storativity (Sy/S)	0.019 to 0.028	0.06 to 0.1	1.30 x10^{-4} to 5.31 X 10^{-5}
Suitability for drinking/ irrigation	Suitable for both (except nitrate and fluoride-affected villages for drinking)		

Table 2.
Aquifer characteristics of Nashik district.

Based on groundwater exploration, depth to water levels, groundwater quality, and aquifer-wise characteristics are given in **Table 2** and the major aquifer map is presented in **Figure 3**.

5. Groundwater resources

5.1 Groundwater resources aquifer-I

Central Ground Water Board and Ground Water Survey and Development Agency (GSDA) have jointly estimated the groundwater resources of Nashik district based on GEC-2015 methodology [5]. The block-wise groundwater resources map is shown in **Figure 21**.

Groundwater resources estimation was carried out for 15,530 sq. km. area out of which 1650.24 sq. km. is under the command and 11838.32 sq. km. is under non-command, the net annual groundwater availability is assessed as 1849.91 million cubic meters (MCM). The gross draft for all uses is estimated at 1081.66 MCM with the irrigation sector being the major consumer having a draft of 1045.72 MCM. The domestic and industrial water requirements are worked out at 35.94 MCM. The net groundwater availability for future irrigation is estimated at 840.45 MCM (**Table 3**). Out of 15 talukas, 3 talukas are semi-critical, 6 are critical and the remaining 9 talukas are safe *(In India, a taluka is the administrative unit below the district).*

5.2 Groundwater resources – aquifer-II

The groundwater resources of the deeper aquifer of the Nashik district were also assessed based on the GEC-2015 methodology. For assessing the resources of Aquifer-II the taluka-wise average thickness of fractured rocks and area occurring under that

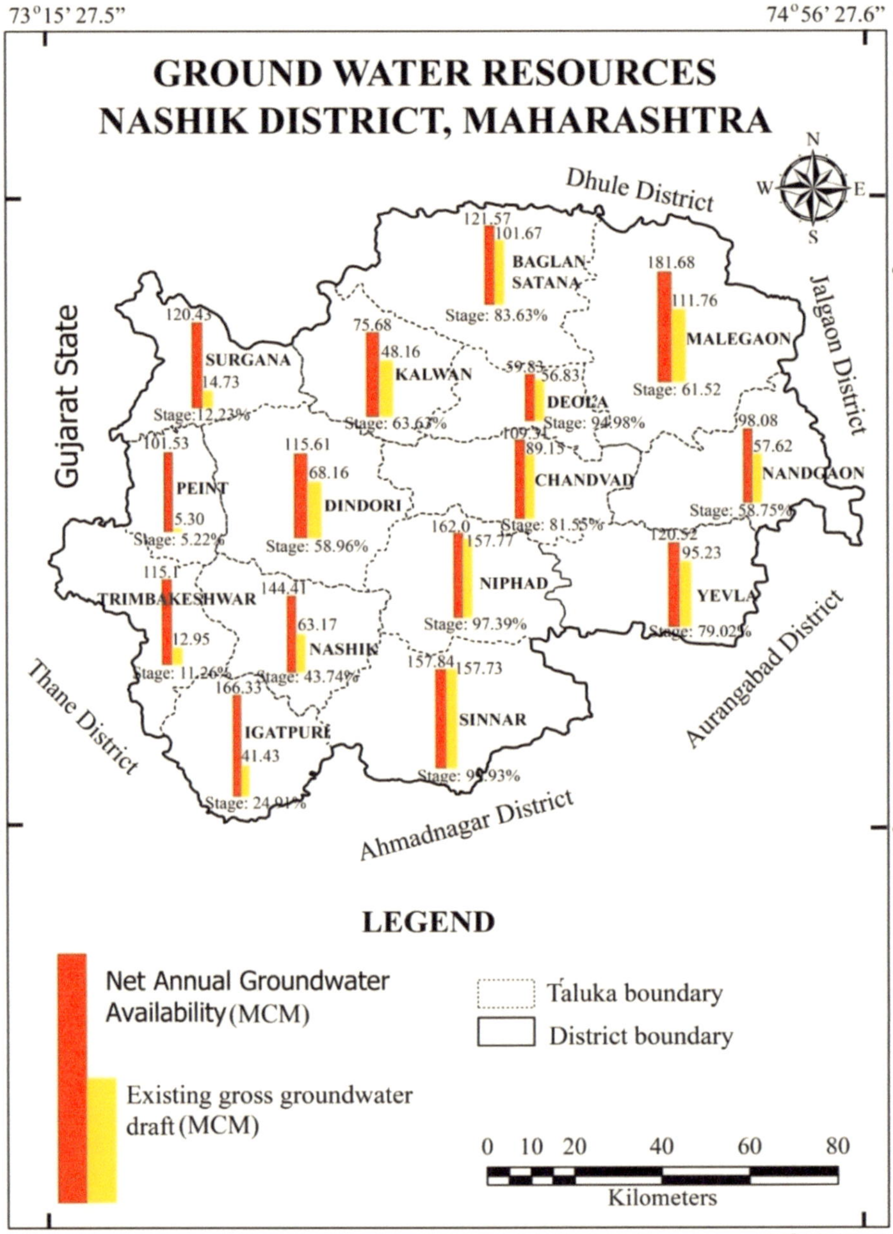

Figure 21.
Groundwater resources (2020), ashik district.

thickness was deduced from **Figure 5**. The specific yield (Sy) and storativity (S) values were taken from CGWB pumping tests data, whereas piezometric water level data was generated from exploration data (**Table 4**). The GEC-2015 methodology suggests the estimation of resources available within the confining layer, as well as those available above the confining layer under hydrostatic pressure [5].

Administrative Unit	Net Annual Groundwater Availability	Existing Gross Groundwater Draft for irrigation	Existing Gross Groundwater Draft for domestic and industrial water supply	Existing Gross Groundwater Draft for All uses	Provision for domestic & industrial requirement supply to 2025	Net Groundwater Availability for future irrigation development	Stage of Groundwater Development %	Category
Baglan Satana	121.57	99.05	2.62	101.67	2.62	26.76	83.63	Semi-Critical
Chandwad	109.31	87.32	1.83	89.15	1.83	22.52	81.55	Semi-Critical
Deola	59.83	54.90	1.93	56.83	1.93	8.70	94.98	Critical
Dindori	115.61	65.61	2.55	68.16	2.55	48.23	58.96	Safe
Igatpuri	166.33	39.47	1.97	41.43	1.97	124.89	24.91	Safe
Kalwan	75.68	46.41	1.75	48.16	1.74	27.52	63.63	Safe
Malegaon	181.68	108.16	3.61	111.76	3.61	78.05	61.52	Safe
Nandgaon	98.08	55.30	2.32	57.62	2.32	40.76	58.75	Safe
Nasik	144.41	61.20	1.97	63.17	1.97	82.90	43.74	Safe
Niphad	162.00	154.02	3.75	157.77	3.75	25.10	97.39	Critical
Peth	101.53	3.55	1.75	5.30	1.75	96.23	5.22	Safe
Sinnar	157.84	154.78	2.95	157.73	2.95	22.57	99.93	Critical
Surgana	120.43	12.63	2.10	14.73	2.10	105.70	12.23	Safe
Trambakeshwar	115.10	10.71	2.25	12.95	2.25	102.14	11.26	Safe
Yeola	120.52	92.63	2.60	95.23	2.60	28.36	79.02	Semi-Critical
Total	**1849.91**	**1045.72**	**35.94**	**1081.66**	**35.94**	**840.45**	**58.45**	

Table 3.
Groundwater resources, aquifer-I (shallow aquifer), Nashik district (2020).

Taluka	Mean thickness of fractured rocks (m)	Area (Sq.km.)	Sy	S	Piezometric head above confining layer (m)	Resource in confining layer (MCM)	Resource above confining layer(MCM)	Total resource (MCM)
1	2	3	4	5	6	7 (= 2*3*4)	8 (= 3*5*6)	9 (= 7 + 8)
Baglan	0.25	440.00	0.005	0.00003	25	0.55	0.33	0.88
Baglan	0.75	425.26	0.009	0.000027	40	2.87	0.46	3.33
Baglan	2.00	463.41	0.007	0.0000012	35	6.49	0.02	6.51
Baglan Total		**1328.67**				**9.91**	**0.81**	**10.72**
Chandvad	10.50	18.78	0.01	0.000057	35	1.97	0.04	2.01
Chandvad	7.50	110.85	0.01	0.000057	35	8.31	0.22	8.53
Chandvad	0.75	101.47	0.009	0.000027	40	0.68	0.11	0.79
Chandvad	2.00	392.75	0.007	0.0000012	35	5.50	0.02	5.51
Chandvad	4.50	173.37	0.008	0.00001	20	6.24	0.03	6.28
Chandvad Total		**797.21**				**22.71**	**0.42**	**23.13**
Deola	7.50	79.50	0.01	0.000057	35	5.96	0.16	6.12
Deola	0.75	139.36	0.009	0.000027	40	0.94	0.15	1.09
Deola	2.00	243.04	0.007	0.0000012	35	3.40	0.01	3.41
Deola	4.50	47.51	0.008	0.00001	20	1.71	0.01	1.72
Deola Total		**509.40**				**12.02**	**0.33**	**12.34**
Dindori	0.75	275.83	0.009	0.000027	40	1.86	0.30	2.16
Dindori	2.00	620.30	0.007	0.0000012	35	8.68	0.03	8.71
Dindori	4.50	7.14	0.008	0.00001	20	0.26	0.00	0.26
Dindori Total		**903.27**				**10.80**	**0.33**	**11.13**
Igatpuri	0.75	345.95	0.009	0.000027	40	2.34	0.37	2.71
Igatpuri	2.00	316.72	0.007	0.0000012	35	4.43	0.01	4.45

Taluka	Mean thickness of fractured rocks (m)	Area (Sq.km.)	Sy	S	Piezometric head above confining layer (m)	Resource in confining layer (MCM)	Resource above confining layer(MCM)	Total resource (MCM)
1	2	3	4	5	6	7 (= 2*3*4)	8 (= 3*5*6)	9 (= 7 + 8)
Igatpuri Total		**662.67**				**6.77**	**0.39**	**7.16**
Kalwan	7.50	25.98	0.01	0.000057	35	1.95	0.05	2.00
Kalwan	2.00	323.67	0.007	0.0000012	35	4.53	0.01	4.54
Kalwan	4.50	261.20	0.008	0.00001	20	9.40	0.05	9.46
Kalwan Total		**610.85**				**15.88**	**0.12**	**16.00**
Malegaon	0.25	464.51	0.005	0.00003	25	0.58	0.35	0.93
Malegaon	0.75	435.54	0.009	0.000027	40	2.94	0.47	3.41
Malegaon	2.00	685.88	0.007	0.0000012	35	9.60	0.03	9.63
Malegaon Total		**1585.93**				**13.12**	**0.85**	**13.97**
Nandgaon	0.75	25.03	0.009	0.000027	40	0.17	0.03	0.20
Nandgaon	2.00	1059.39	0.007	0.0000012	35	14.83	0.04	14.88
Nandgaon Total		**1084.42**				**15.00**	**0.07**	**15.07**
Nashik	0.75	326.31	0.009	0.000027	40	2.20	0.35	2.56
Nashik	2.00	411.02	0.007	0.0000012	35	5.75	0.02	5.77
Nashik Total		**737.34**				**7.96**	**0.37**	**8.33**
Niphad	10.50	68.62	0.01	0.000057	35	7.21	0.14	7.34
Niphad	7.50	127.46	0.01	0.000057	35	9.56	0.25	9.81
Niphad	0.25	65.19	0.005	0.00003	25	0.08	0.05	0.13
Niphad	0.75	175.28	0.009	0.000027	40	1.18	0.19	1.37
Niphad	2.00	449.20	0.007	0.0000012	35	6.29	0.02	6.31
Niphad	4.50	243.83	0.008	0.00001	20	8.78	0.05	8.83

Taluka	Mean thickness of fractured rocks (m)	Area (Sq.km.)	Sy	S	Piezometric head above confining layer (m)	Resource in confining layer (MCM)	Resource above confining layer(MCM)	Total resource (MCM)
1	2	3	4	5	6	7 (= 2*3*4)	8 (= 3*5*6)	9 (= 7 + 8)
Niphad Total		**1129.57**				**33.10**	**0.70**	**33.79**
Peint Total	**0.75**	**496.47**	**0.009**	**0.000027**	**40**	**3.35**	**0.54**	**3.89**
Sinnar	7.50	38.61	0.01	0.000057	35	2.90	0.08	2.97
Sinnar	0.25	142.22	0.005	0.00003	25	0.18	0.11	0.28
Sinnar	0.75	557.30	0.009	0.000027	40	3.76	0.60	4.36
Sinnar	2.00	403.00	0.007	0.0000012	35	5.64	0.02	5.66
Sinnar	4.50	126.82	0.008	0.00001	20	4.57	0.03	4.59
Sinnar Total		**1267.96**				**17.04**	**0.83**	**17.87**
Surgana	0.75	355.74	0.009	0.000027	40	2.40	0.38	2.79
Surgana	2.00	367.43	0.007	0.0000012	35	5.14	0.02	5.16
Surgana Total		**723.17**				**7.55**	**0.40**	**7.94**
Trimbakesh-war	0.75	424.76	0.009	0.000027	40	2.87	0.46	3.33
Trimbakesh-war	2.00	256.65	0.007	0.0000012	35	3.59	0.01	3.60
Trimbakesh-war Total		**681.41**				**6.46**	**0.47**	**6.93**
Yevla	0.75	563.22	0.009	0.000027	40	3.80	0.61	4.41
Yevla	2.00	406.05	0.007	0.0000012	35	5.68	0.02	5.70
Yevla	4.50	0.25	0.008	0.00001	20	0.01	0.00	0.01
Yevla Total		**969.51**				**9.50**	**0.63**	**10.12**
Nashik District Total		**25509.68**				**369.47**	**13.30**	**382.77**

Table 4.
Groundwater resource assessment of aquifer – II (deeper aquifer).

As per the estimation, the resources in confining layer are 369.47 MCM, whereas resources above confining layer are assessed as 13.30 MCM and the total groundwater resources come to the tune of 382.77 MCM. The net groundwater resources of Aquifer – I are 1849.90 MCM, whereas that of Aquifer-II are only 382.77 MCM, which is about 21% of dynamic resource availability indicating that the deeper groundwater resources are scarce and more vulnerable to overextraction.

6. Groundwater related issues

6.1 Declining water levels

Pre-monsoon groundwater falling trend of more than 0.2 m/year occurs in 2832.01 sq. km of the area in parts of Peth, Surgana, Baglan, Malegaon, Yevla, Igatpuri, Trambakeshwar, Nashik and Nandgaonblocks. Post-monsoon groundwater falling trend of more than 0.2 m/year occurs in 216.39 sq. km of the area in parts of Malegaon, Igatpuri, Nashik, and Baglan blocks.

6.2 Rainfall and droughts

Based on the long-term rainfall analysis from 1998 to 2021, it is observed that Chandwad, Igatpuri, Malegaon, Nashik, Niphad, Peinth, Surgana, Trimbakeshwar, and Yeola blocks experienced declining rainfall trends. Severe droughts have been observed in Trimbakeshwar, Surgana, Sinnar, Peinth, Nandgaon, Malegaon, Kalwan, Igatpuri, Chandwad, and BaglanSatana Blocks at least once, the rest of the blocks never experienced severe drought conditions; however, it has experienced moderate droughts.

6.3 Overexploitation

Three blocks of the district, that is . Deola, Niphad, and Sinnar fall under the critical category and three blocks fall under semi-critical, that is.BaglanSatana, Chandwad, and Yeola and the rest of the blocks are under the safe category in 2020. In critical and semi-critical blocks, a declining groundwater level trend has been observed both in pre-and post-monsoon seasons.

6.4 Low groundwater potential

Groundwater potential areas have been identified in 14,731 sq. km ($\approx$ 90%) in the Nashik district, where yield remains less than 15 m^3/day, mostly due to limited depth of weathering and fractures in Aquifer-I (Basalt). Limited aquifer potential of Aquifer-II (Basalt) is observed in about 9390 sq. km ($\approx$59%) area of the district (Yield potential $<$1.0 LPS). The sustainability of both Aquifers- I & II is, thus, low and the wells normally sustain pumping between less than 1 and 3 hours.

6.5 Cultivation of cash crops

Various cash crops, *viz.*, grapes, onion, pomegranate, and sugarcane, are cultivated in the district, and the major irrigation source is groundwater. This has led to the severe exploitation of groundwater resources in 6 talukas. The water requirement in these crops is also higher as compared to other traditional crops, thus laying more stress on groundwater. Although micro-irrigation techniques are practiced in most areas for the cultivation of grapes, other cash crops are also required to be brought under the ambit of micro-irrigation.

7. Groundwater management plan

The management plan has been proposed to effectively manage groundwater resources and to arrest overexploitation [6]. The management plan comprises two components namely supply-side management and demand-side management. The supply-side management is based on surplus surface water availability and the unsaturated thickness of the aquifer, whereas the demand-side management is by the use of micro-irrigation techniques and changes in cropping pattern. In addition to this, in some parts of the district, groundwater development [7] is also formulated and recommended to authorities for implementation.

7.1 Supply-side management

The supply-side management of groundwater resources can be done through the artificial recharge of surplus runoff available within river subbasins and micro watersheds [8, 9]. Also, it is necessary to understand the unsaturated aquifer volume available for recharge. The unsaturated volume of the aquifer was computed based on the area feasible for recharge, the unsaturated depth < 3 m bgl, and the specific yield of the aquifer. **Table 5** gives the block-wise supply-side interventions through recharge structures.

The area identified for recharge measures in the district is worked out at 6334.28 sq.km.; however, in 5 talukas namely Igatpuri, Nasik, Peth, Surgana, and Trambakeshwar, there is no scope for aquifer recharge due to shallow post-monsoon water levels and rising water level trends. The total unsaturated volume available for artificial recharge is 25440.80 MCM, which ranges from 42.64 MCM in Dindorito to 8565.54 MCM in BaglanSatanataluka. Considering 75% efficiency the surface water required will be 33836.28 MCM, whereas the surface water available in the district is only 265.41 MCM, thus there is a huge shortfall of the surface water required for recharging the aquifers. The available surplus surface runoff water can be utilized for aquifer recharge through the construction of percolation tanks, check dams, gabion, underground bandharas, Nala bunds, continues contour trenches (CCT), and other site-specific structures; however, to arrive at specific numbers of the most feasible recharge structures, that is. check dams and percolation tanks were considered. The check dams of 10 TCM capacities with three fillings in a year and percolation tanks of 100 thousand cubic meters (TCM) capacity with two fillings in a year were considered to arrive at the number of the structures. Thus, a total of 649 percolation tanks and 1857 check dams are proposed with the available surface water resources. The tentative locations of these structures are shown in **Figure 22**.

7.2 Demand-side management

As seen above, supply-side interventions have limitations due to the limited availability of water sources, thereby hindering proper management of the aquifer

DOI: http://dx.doi.org/10.5772/intechopen.111644

Taluka	Geographical Area	Area identified for AR (sq. km.)	Unsaturated Volume (MCM)	Surplus surface water required for AR (MCM)	Surplus surface water available for AR (MCM)	Proposed number of structures		Total Volume of Water expected to be recharged@ 75% efficiency (MCM)		Total recharge @ 75% efficiency (MCM)
						PT	CD	PT	CD	
Baglan Satana	1481.65	883.35	8565.54	11392.17	37.01	97	278	14.58	6.25	20.84
Chandwad	890.07	692.59	1334.27	1774.58	29.02	76	218	11.44	4.90	16.34
Deola	577.46	294.13	2554.73	3397.79	12.32	32	93	4.86	2.08	6.94
Dindori	1141.93	38.88	42.64	56.71	1.63	3	9	0.45	0.19	0.64
Igatpuri	932.97	0	0	0	0.00	0	0	0.00	0.00	0.00
Kalwan	702.96	577.97	4747.58	6314.28	24.22	64	182	9.54	4.09	13.63
Malegaon	1806.39	1186.19	1542.05	2050.92	49.70	108	308	16.19	6.94	23.13
Nandgaon	1164.88	785.76	903.63	1201.82	32.92	63	181	9.49	4.07	13.55
Nasik	846.31	0	0	0	0.00	0	0	0.00	0.00	0.00
Niphad	1151.75	699.99	2187.14	2908.89	29.33	77	220	11.56	4.95	16.51
Peth	715.87	0	0	0	0.00	0	0	0.00	0.00	0.00
Sinnar	1326.48	704.41	2658.77	3536.16	29.51	77	222	11.63	4.98	16.61
Surgana	912.15	0	0	0	0.00	0	0	0.00	0.00	0.00
Trambakeshwar	873.34	0	0	0	0.00	0	0	0.00	0.00	0.00
Yeola	1005.79	471.01	904.46	1202.93	19.74	52	148	7.78	3.33	11.11
Total	**15,530**	**6334.28**	**25440.80**	**33836.28**	**265.41**	**649**	**1857**	**97.51**	**41.79**	**139.30**

Table 5.
Proposed supply-side intervention through aquifer recharge.

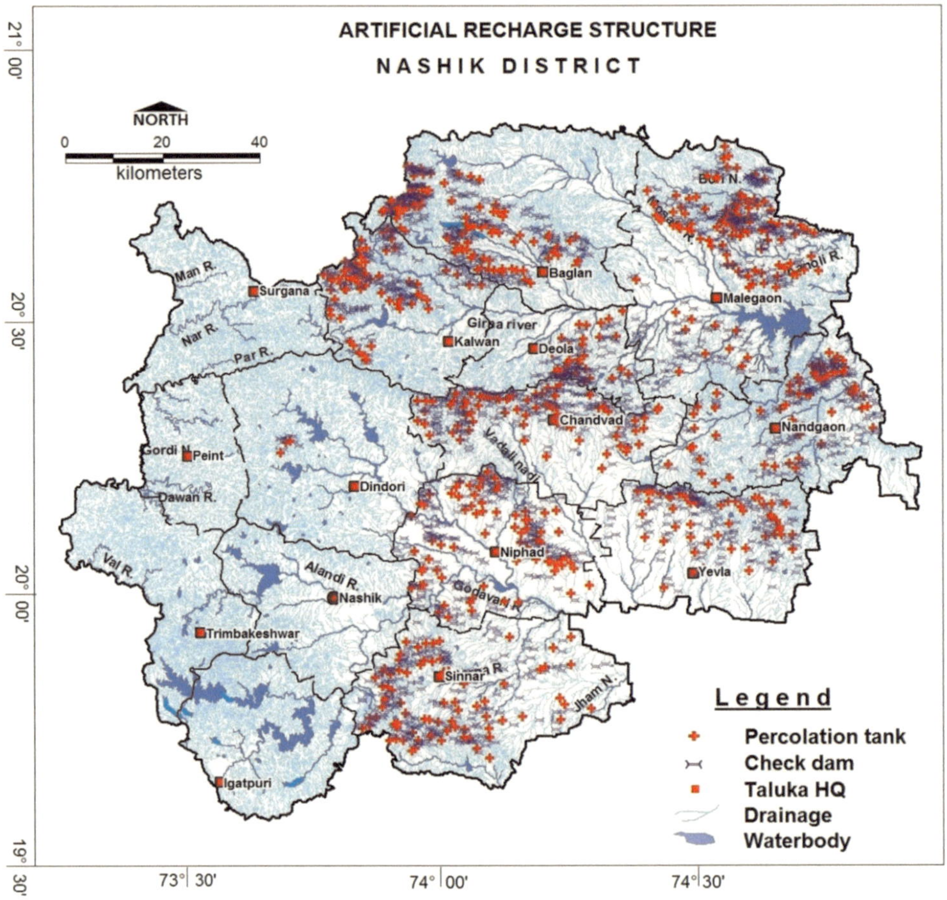

Figure 22.
Location of proposed aquifer recharge structures.

system. In such cases, it becomes imperative to adopt demand-side interventions and reduce our demand for groundwater by adopting micro-irrigation techniques. Generally speaking, traditional watering methods can lose as much as 50% or even more of the water applied as evaporation and infiltration losses. Whereas, the application of a micro-irrigation system (MIS) can increase yields and decrease water use, fertilizer quantity, and labor requirements [10]. In addition, other indirect benefits, such as significant energy savings, are observed, which are associated with the electricity required to pump water from the aquifer. The MIS system operates under low pressure, and according to the precise water requirement of the crop. Each dripper/emitter supplies a precisely controlled quantity of water and nutrients directly to the root zone of the plant [9, 10].

The demand-side management is proposed in areas where the stage of groundwater development is relatively high and adopting micro-irrigation techniques for water-intensive crops or change in cropping pattern or both are required to save water [9]. In Nasik district, two major cash and water-intensive crops are proposed to be brought under drip irrigation, that is, sugarcane and onion. For sugarcane, the crop water

Taluka	Geographical Area (Sqkm)	Area proposed to be covered under Drip (sq. Km)		Volume of Water expected to be saved (MCM). WUE- 0.57 m for sugarcane, 0.26 m for onion		Total water saved (MCM)
		Sugarcane crop area	Onion crop area			
Baglan Satana	1481.65	19.03	50.00	10.85	13.00	23.85
Chandwad	890.07	0.11	100.00	0.06	26.00	26.06
Deola	577.46	1.19	100.00	0.68	26.00	26.68
Dindori	1141.93	10.00	10.00	5.70	2.60	8.30
Igatpuri	932.97	2.75	0.00	1.57	0.00	1.57
Kalwan	702.96	7.25	50.00	4.13	13.00	17.13
Malegaon	1806.39	3.58	100.00	2.04	26.00	28.04
Nandgaon	1164.88	2.32	50.00	1.32	13.00	14.32
Nasik	846.31	12.77	0.00	7.28	0.00	7.28
Niphad	1151.75	40.00	100.00	22.80	26.00	48.80
Peth	715.87	0.00	0.00	0.00	0.00	0.00
Sinnar	1326.48	15.00	130.00	8.55	33.80	42.35
Surgana	912.15	0.00	0.00	0.00	0.00	0.00
Trambakeshwar	873.34	0.00	0.00	0.00	0.00	0.00
Yeola	1005.79	2.84	100.00	1.62	26.00	27.62
Total	**15,530**	**116.84**	**790.00**	**66.60**	**205.40**	**272.00**

Table 6.
Area proposed for drip irrigation demand side management.

requirement by traditional flooding method is 2.45 m, and by drip irrigation, it is 1.88 m, thus a savings of 0.57 m (23%) can be achieved. Similarly, in the case of onion, the crop water requirement by the traditional flooding method is 0.78 m, and by drip irrigation, it is 0.52 m, thus a savings of 0.26 m (33%) can be achieved. Based on these calculations, the taluka wise demand-side management plan suggested for Nashik district is outlined in **Table 6**.

In Nashik district, an area of 116.84 sq. km. of groundwater irrigated sugarcane crop, and 790 sq. km. of onion crop is proposed to be brought under drip irrigation. The volume of water expected to be saved is estimated at 66.60 MCM for sugarcane and 205.40 MCM for onion, totaling 272 MCM of groundwater savings. The areas of sugarcane and onion, which can be brought under drip irrigation are plotted on the map and presented in **Figure 23**. If we compare, the groundwater augmentation by aquifer recharge measures is 139.30 MCM; however, water savings by micro-irrigation are much higher at 272 MCM. Thus, demand-side interventions are more beneficial, economical, and suitable for areas where there is limited source water availability for recharge and areas with high groundwater consumption for irrigation, especially cash crops and water-intensivecrops (**Figure 24**).

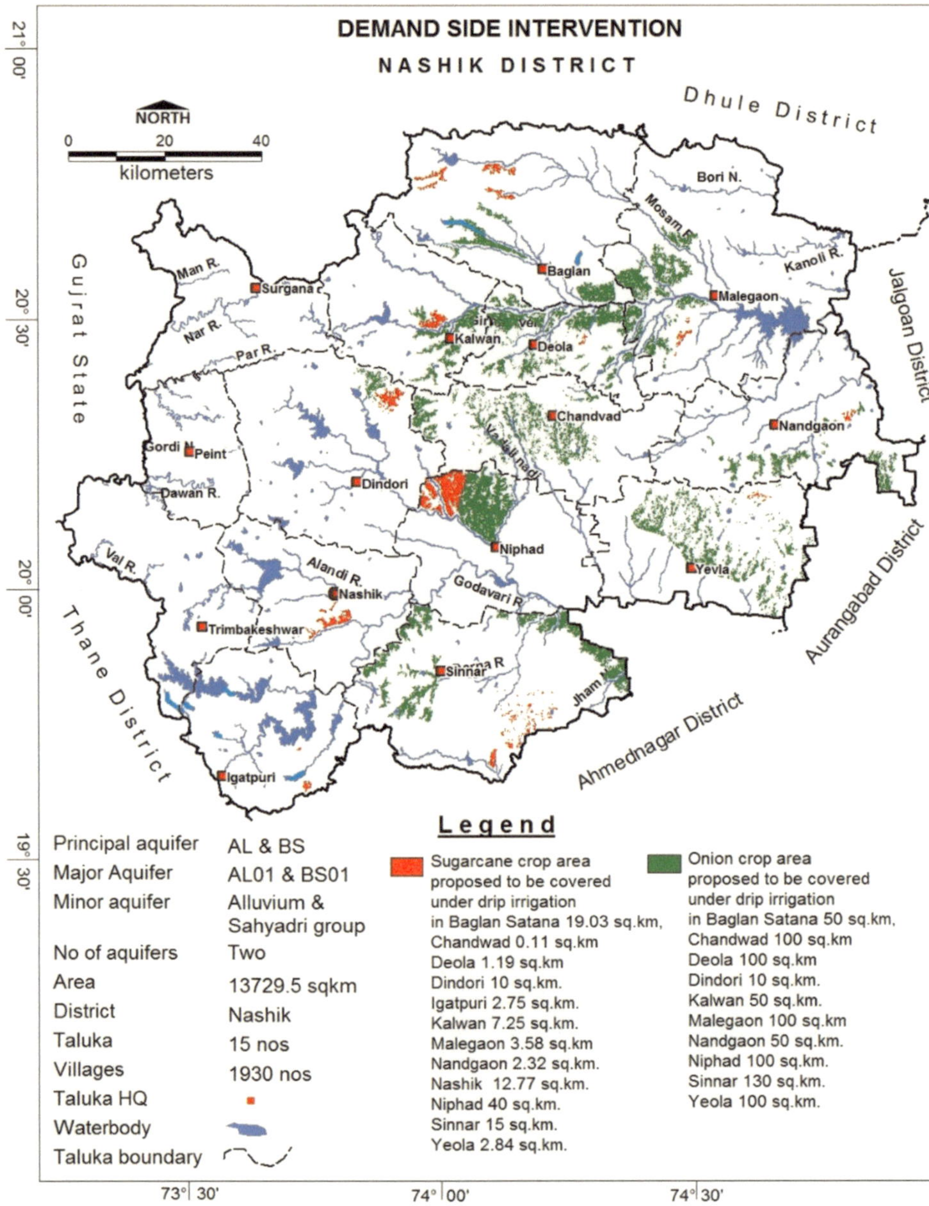

Figure 23.
Areas proposed for demand-side interventions.

7.3 Expected benefits of aquifer management plan

The impact of groundwater management plans on the groundwater system in the district after its implementation is evaluated, and the outcome shows significant improvement in the groundwater scenario in all blocks (**Table 7**).

The total groundwater resource available after supply-side interventions is 1989.21 MCM, whereas the total groundwater draft after the demand-side intervention is

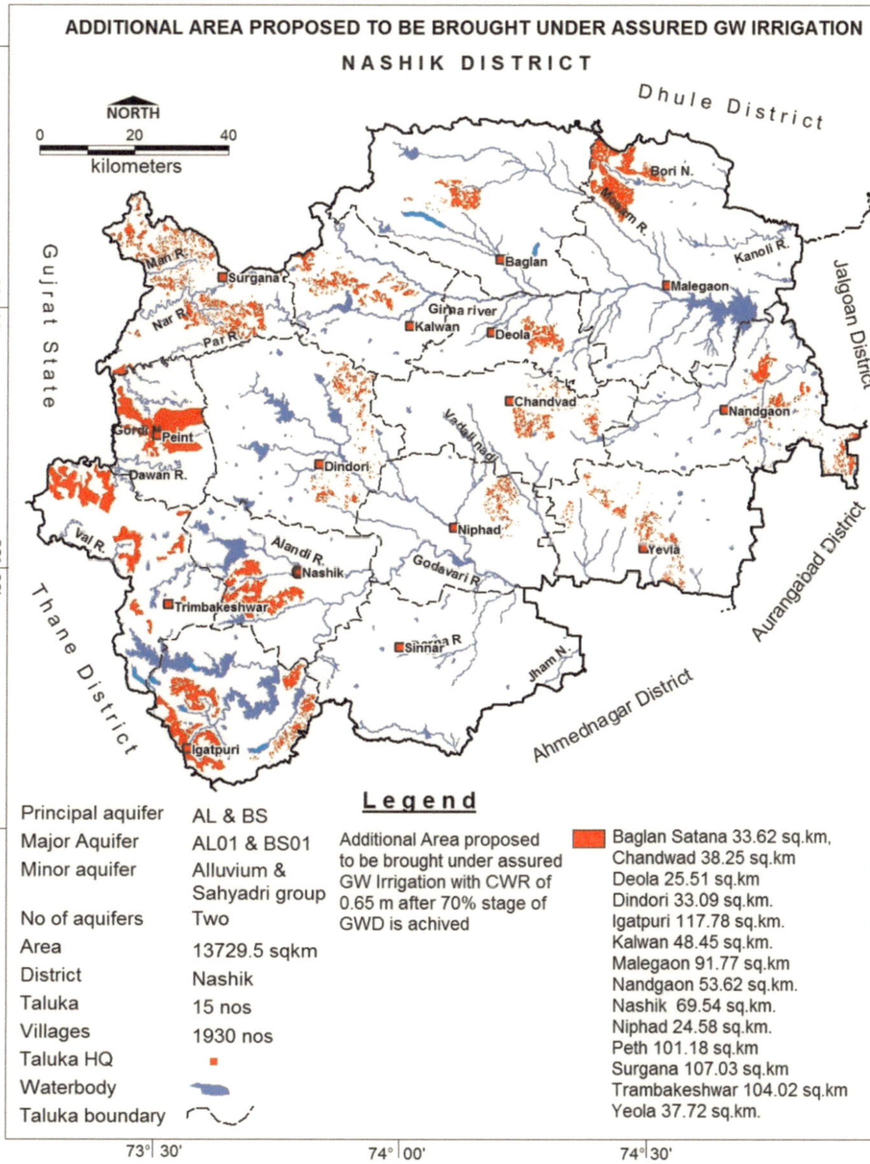

Figure 24.
Groundwater savings (MCM) due to supply and demand side interventions.

809.66 MCM. Thus, about 582.79 MCM of groundwater is available, which can improve the stage of groundwater development by 17% from 58% to around 41%. With this, an additional area of 896.60 sq. km can be irrigated. The tentative locations of these areas are shown in **Figure 25**.

7.4 Groundwater development plan

As per groundwater resource assessment data of 2020, balance groundwater resources to the tune of 582.79 MCM are available for development ranging from 6.74

Block	Water expected to be recharged/ conserved by supply-side interventions (MCM)	Groundwater resources after supply-side interventions (MCM)	Improvement in Stage of GWD after supply-side interventions (%)	Water expected to be saved by Demand side interventions (MCM)	Groundwater Extraction after demand-side intervention	Improvement in Stage of GWD after supply-side and demand side interventions (%)	Balance GWR available for GW Development to enhance the stage of GWD is brought to 70% (MCM)	Additional Area (sq. km.) proposed to be brought under assured GW irrigation with av. CWR of 0.65 m after 70% stage of GWD is achieved (sq. Km)
Baglan Satana	20.84	142.41	71.39	23.85	77.82	54.65	21.86	33.63
Chandwad	16.34	125.65	70.95	26.06	63.08	50.21	24.87	38.26
Deola	6.94	66.77	85.11	26.68	30.15	45.16	16.59	25.52
Dindori	0.64	116.24	58.63	8.30	59.86	51.49	21.51	33.10
Igatpuri	0.00	166.33	24.91	1.57	39.87	23.97	76.56	117.79
Kalwan	13.63	89.31	53.92	17.13	31.02	34.74	31.49	48.45
Malegaon	23.13	204.81	54.57	28.04	83.72	40.88	59.65	91.76
Nandgaon	13.55	111.64	51.61	14.32	43.30	38.78	34.85	53.61
Nasik	0.00	144.41	43.74	7.28	55.89	38.70	45.20	69.54
Niphad	16.51	178.51	88.38	48.80	108.97	61.04	15.99	24.60
Peth	0.00	101.53	5.22	0.00	5.30	5.22	65.77	101.18
Sinnar	16.61	174.45	90.41	42.35	115.38	66.14	6.74	10.37
Surgana	0.00	120.43	12.23	0.00	14.73	12.23	69.57	107.03
Trambakeshwar	0.00	115.10	11.26	0.00	12.95	11.26	67.61	104.02
Yeola	11.11	131.62	72.35	27.62	67.61	51.37	24.53	37.73
Total	**139.30**	**1989.21**	**54.38**	**272.00**	**809.66**	—	**582.79**	**896.60**

Table 7.
Expected benefits after management options.

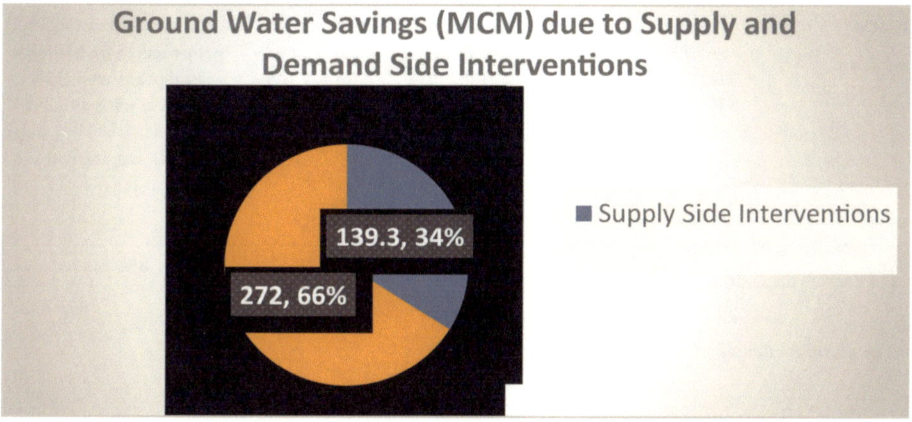

Figure 25.
Additional area proposed to be bought under assured GW irrigation.

MCM in Sinnartaluka to 76.56 MCM in Igatpuritaluka. While proposing a groundwater development plan, it is imperative to consider the shallow aquifer for major groundwater extraction as it gets annually replenished, as well as it is the most promising, sustainable aquifer in the district. Therefore, 90% of these resources are proposed to be developed by 34,967 dug wells tapping shallow Aquifer-I, and the remaining 10% are proposed to be developed by 5828 borewells tapping moderate to deeper Aquifer-II. Thus, about 34,967 dug wells and 5828 borewells can be constructed. The block-wise details are given in **Table 8**.

Block	Balance GWR available for GW Development after STAGE OF GWD is brought to 70% (MCM)	Proposed No. of DW @1.5 ham for 90% of GWR Available)	Proposed No. of BW @1.0 ham for 10% of GWR Available)	Additional Area (sq.km.) proposed to be brought under assured GW irrigation with av. CWR of 0.65 m after 70% stage of GWD is achieved (sq. Km)
Baglan Satana	21.86	1312	219	33.63
Chandwad	24.87	1492	249	38.26
Deola	16.59	995	166	25.52
Dindori	21.51	1291	215	33.10
Igatpuri	76.56	4594	766	117.79
Kalwan	31.49	1890	315	48.45
Malegaon	59.65	3579	596	91.76
Nandgaon	34.85	2091	348	53.61
Nasik	45.20	2712	452	69.54
Niphad	15.99	959	160	24.60
Peth	65.77	3946	658	101.18
Sinnar	6.74	404	67	10.37
Surgana	69.57	4174	696	107.03
Trambakeshwar	67.61	4057	676	104.02

Block	Balance GWR available for GW Development after STAGE OF GWD is brought to 70% (MCM)	Proposed No. of DW @1.5 ham for 90% of GWR Available)	Proposed No. of BW @1.0 ham for 10% of GWR Available)	Additional Area (sq.km.) proposed to be brought under assured GW irrigation with av. CWR of 0.65 m after 70% stage of GWD is achieved (sq. Km)
Yeola	24.53	1472	245	37.73
Total	**582.79**	**34,967**	**5828**	**896.60**

Table 8.
Blockwise groundwater development plan.

8. Conclusions

The correlation of the yield potential of Aquifer-I with the depth of occurrence and weathered/fractured thickness indicates that even though the aquifer occurs down to the moderate depth of 10–15 m with a sufficient thickness of 8 to 12 m it does not guarantee adequate yield in hard rock basaltic aquifer. Similarly, the correlation of the yield potential of Aquifer-II with the depth of occurrence and fractured thickness also indicates that even though the aquifer occurs down to moderate to deep depths of 80 to 140 m in a major part of the area, the yield is not dependent on the depth of occurrence of the aquifer. Thus, the common tendency of the borewell culture of going deep for getting more groundwater does not hold in hard rock areas, especially in basaltic aquifers.

The deeper water level of more than 50 m bgl is observed in major parts of the district covering an area of 7097.14 Sq.km. This may be due to the overexploitation of groundwater and it also indicates that the shallow aquifer is not able to sustain the demand; hence, groundwater resources are also being withdrawn from deeper aquifers in this area. The net groundwater resources of the aquifer are 1849.90 MCM, whereas that of Aquifer-II are only 382.77 MCM, which is about 21% of dynamic resource availability indicating that the deeper groundwater resources are scarce and more vulnerable to overextraction. Devising a groundwater management strategy or plan for hard rock aquifers is always challenging [7] as compared to alluvium aquifers, which occur in the vast expanses of major alluvial basins of India. This is due to the highly diversified occurrence and considerable variations in the hard rock groundwater availability. The problem is further compounded as in this case, the area has witnessed droughts, declining water levels, and low yield potentials. The multi-pronged aquifer management plan is suggested for the area, that is. supply-side (augmentation), demand-side (savings), and development side to resolve the issues.

The supply-side interventions, that is. recharge measures in the district are feasible in 6334.28 sq.km.; however in 5 talukas namely Igatpuri, Nasik, Peth, Surgana, and Trambakeshwar, there is no scope for aquifer recharge due to shallow post-monsoon water levels and rising water level trends. However, in patches, these talukas also face water scarcity in the peak summer season. This dichotomy of availability and nonavailability of groundwater concerning the season is very common in hard rock aquifers. The total noncommitted surface runoff required for recharging the aquifers in feasible areas of 6334 sq. km is 33836.28 MCM, whereas the surface water available in the district is only 265.41 MCM, thus there is a huge shortfall of the surface water required for recharging the aquifers. The available surplus surface runoff water can be utilized for aquifer recharge through the construction of 649 percolation tanks, 1857 check dams/gabion, and various other site-specific structures such as underground bandharas, Nala bunds, and CCT (continuous contour trenching).

The demand side interventions, that is., water saving measures are proposed in the areas with the high level of the stage of extraction, an area of 116.84 sq. km. of groundwater irrigated sugarcane crop and 790 sq. km. of onion crop is proposed to be brought under drip irrigation. The volume of water expected to be saved is estimated at 66.60 MCM for sugarcane and 205.40 MCM for onion, totaling 272 MCM of groundwater savings.

If we compare, the supply-side interventions by aquifer recharge measures can augment resources to the tune of 139.30 MCM; however, demand-side interventions by micro-irrigation are much higher @ 272 MCM. Thus, demand-side interventions are more beneficial, economical, and suitable for areas, where there is limited source water availability for recharge and areas with high groundwater consumption for irrigation, especially cash crops and water-intensive crops. The study has indicated that the supply-side interventions have got apparent limitations due to inadequate availability of surplus/noncommitted surface water, as well as the low storage potential of hard rock aquifers.

With supply-side and demand-side interventions, it is expected that about 411 MCM of groundwater would be available to bring down the overall stage of groundwater development of the district from 58.45 to 47.81%. This plan would also address the issue of overextraction in some talukas such as Baglan Satana, Chandwad, Deola, Niphad, Sinnar, and Yeola, where the stage of extraction can be brought down to 70% (safe category).

The management plan provides a scope for the development of available and additional groundwater resources added/saved to the system through the construction of 34,967 dug wells and 5828 borewells in a phased manner in the selected areas. This will provide assured irrigation to an additional area of 896.60 sq. km. The management plan suggested for the area is holistic and implementable with further fine-tuning/inputs at the local scale. It will also lead to the upliftment of the socioeconomic status of the people, especially of those located in tribal pockets of the district. These interventions also need to be supported by capacity building measures, self-regulation, and an institutional framework for participatory groundwater management.

Author details

Prabhat Jain[1*], Abhay Soni[2] and Rahul Shende[3]

1 Central Ground Water Board, Central Region, Ministry of Jal Shakti, Government of India, Nagpur, Maharashtra, India

2 Nagpur Research Centre, CSIR-Central Institute of Mining and Fuel Research (CSIR-CIMFR), Nagpur, Maharashtra, India

3 Central Ground Water Board, South Western Region, Ministry of Jal Shakti, Government of India, Bengaluru, Karnataka, India

*Address all correspondence to: abhayksoni@gmail.com

References

[1] CGWB. Concept Note on National Project on Aquifer Management (NAQUIM). Central Ground Water Board (CGWB), Govt of India; 2011a. 14 p. Available from: http://cgwb.gov.in/

[2] CGWB. Manual on Aquifer Mapping. Central Ground Water Board (CGWB), Govt of India; 2011b. 72 p. Available from: http://cgwb.gov.in/

[3] CGWB and GSDA. Dynamic Groundwater Resources of Maharashtra & Groundwater Survey and Development Agency (GSDA). Govt. of Maharashtra, Technical Reports; 2020-21. p. 167

[4] CGWB. Aquifer Maps and Groundwater Management Plan, Nashik District, Maharashtra. Central Ground Water Board (CGWB), Govt of India; 2018. 167 p. Available from: http://cgwb.gov.in/

[5] CGWB. The Ground Water Resource Estimation Committee (GEC-2015), Technical Report of Central Ground Water Board (CGWB). Govt of India; 2017. p. 142. Available from: http://cgwb.gov.in/

[6] CGWB. Master Plan for Artificial Recharge to Groundwater in India. Central Ground Water Board (CGWB), Govt of India; 2021. p. 198. Available from: http://cgwb.gov.in/

[7] Limaye SD. Groundwater Development in Hard Rocks Groundwater Vol II, Encyclopedia of Life Support System (EOLSS). p. 6. Available from: http://www.eolss.net/Eolss-sampleAllChapter.aspx. [Accessed: November 14, 2022]

[8] CGWB. Manual on Artificial Recharge of Groundwater. Central Ground Water Board (CGWB), Govt of India; 2007. p. 185. Available from: http://cgwb.gov.in/

[9] Singh SP, Singh B. Water resource Management in a Hard Rock Terrain- a Case Study of Jharkhand state, India. APCBEE Procedia. 2012;**1**:245-251, (ISSN 2212-6708. DOI: 10.1016/j.apcbee.2012.03.040), 10.1016/j.apcbee.2012.03.040)

[10] Prabhat J. Micro Irrigation – A Smart Technique for Sustainable Agriculture, Souvenir. Nagpur: Engineers Forum; 2022. p. 3

Chapter 7

Groundwater Dynamics in Transboundary Aquifers of Southern Africa

Malebo Matlala

Abstract

Groundwater resources are indispensable not only in water scarce or water stressed countries, but globally as a dependable reservoir and an alternative resource of freshwater. This study assessed the spatio-temporal variability of groundwater resources within two of the biggest transboundary aquifers that South Africa shares with its neighbouring countries. Groundwater dynamics in the Karoo-Sedimentary Transboundary Aquifer (KSTA) as well as the Stampriet Transboundary Aquifer System (STAS) were studied over a period of 72 years from 1948-2020. The study explored the use of historical groundwater storage data acquired through the use of Remote Sensing (RS) techniques, coupled with the use of Geographical Information Systems (GIS) to map spatio-temporal variability in groundwater storage. Groundwater resources of the Karoo-Sedimentary Transboundary Aquifer were found to be declining over time, with an overall decline of just over 5.4 km^3, whereas groundwater resources in the Stampriet remained relatively constant, with an overall increase of 0.2 km^3 over the past 72 years. The results show that RS techniques coupled with GIS applications are invaluable where there is a dearth of scientific data and information, furthermore, their use in the monitoring, management and protection of groundwater resources can be applicable on the local, regional and international scales.

Keywords: transboundary aquifer, southern Africa, groundwater, aquifer monitoring, Stampriet transboundary aquifer, Karoo sedimentary transboundary aquifer

1. Introduction

Water is a valuable commodity, not only for the human population, but also for all living as well as non-living biota. The availability of freshwater resources coupled with the quality of the available resource determines not only where life settles, but the quality of that life too [1].

Although nearly 70% of the surface of the earth is covered with water, over 90% of this is ocean water; an additional 2.2% is contained in ice sheets, glaciers and ice caps [2]. Therefore, according to Davies and Day [2], the fraction of water suitable for domestic use (potable water) is less than 1% of global water supplies. Furthermore, this proportion of freshwater is made up of atmospheric water, surface as well as groundwater resources [2]. With such limited availability, freshwater resources are not only reserved for domestic use and the maintenance of aquatic ecosystems, benefits derived from available freshwater resources also include agricultural use, industrial production, mining as well as navigation, thus enabling and facilitating global socio-economic development. For this reason alone, global water security is of paramount concern.

Freshwater resources are, however, disproportionately distributed, where some areas have sufficient supplies of the resource, whereas the resource is gravely limited in other areas, thus threatening not only local or regional water security, but socio-economic advancement too. Moreover, freshwater resources are not only threatened by their natural distribution and availability, population growth coupled with economic expansion also increases the demand for freshwater resources worldwide. In addition, climate variability also threatens the supply and renewal of freshwater resources, more so in water-scarce areas, where extreme effects of climate change, such as droughts, will further exacerbate water scarcity. Although finite, freshwater resources are renewed through precipitation, which, similar to the distribution of freshwater supplies, is also disproportionately distributed. Thus, in order to protect the resource, it is important to safeguard the use of both surface and groundwater resources.

Groundwater resources are often utilised where surface water resources are limited, augmenting surface water supplies. Not only does groundwater supplement surface waters in arid and semi-arid areas, groundwater resources also sustain associated or hydraulically linked ecosystems through river base flows [3, 4], acting as a significant buffer in times of droughts [3, 5]. Thus, groundwater resources are fundamental to the Earth's supply of freshwater. These subsurface resources, which act as a natural reservoir for freshwater, are important for socio-economic development as well as the alleviation of poverty [6–8].

Groundwater resources provide the most dependable and most easily accessible source of freshwater in most arid and semi-arid areas with limited surface water supplies. Many socio-economic activities are reliant on groundwater resources, and over half the global population also relies on groundwater for their domestic supplies, where over 60% is used for agricultural irrigation, 25% is allocated for domestic use, and 10% is used by industry [9–11]. This allocation, however, differs from region to region, for example, groundwater augments drinking water in many developed countries, whereas, in semi-arid regions of the world, it serves as the main source of water [10]. According to Foster and Chilton [9], groundwater systems represent the world's most predominant freshwater reservoirs. However, in many parts of the world, groundwater resources are excluded in the management of water resources, resulting in the overexploitation of the resource with no regard to its sustainability. Although out of sight, and often not easily quantifiable, groundwater resources should be monitored, conserved and efficiently managed to ensure their equitable and sustainable utilisation on national, regional and global scales; however, their subterranean nature makes this difficult to implement, especially where these resources traverse political borders.

1.1 Groundwater and aquifers

Groundwater is found below the surface of the earth, collecting in spaces in the soil and sediment and also filling up interstices between rocks [12]. These rock formations, together with the water stored within the crevices of the rock, are known as aquifers. Therefore, the shape and size of the spaces within the aquifer (rock formation) affect the total volume of water that can be stored in the spaces, a term referred to as porosity [13]. Moreover, in addition to an aquifer's storage capacity, an

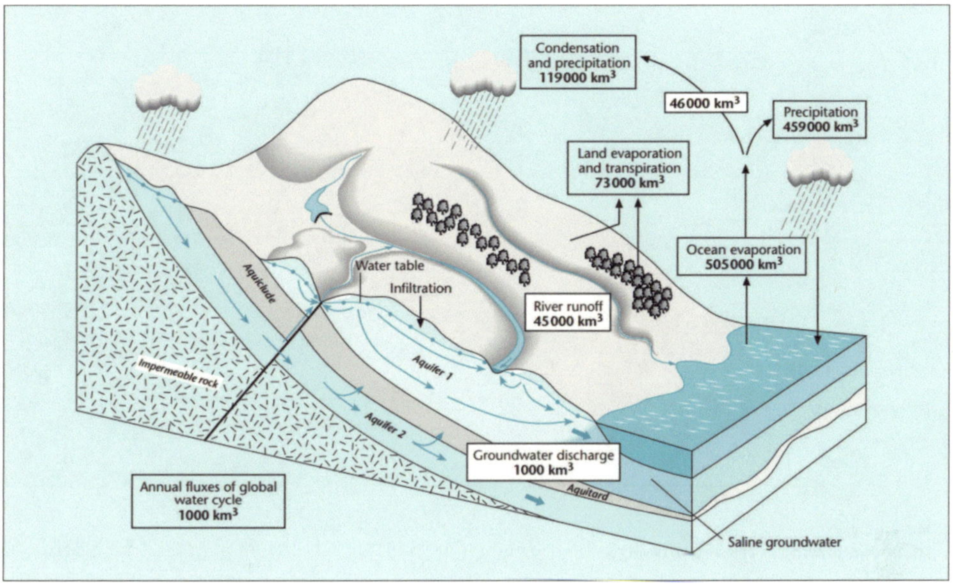

Figure 1.
A representation of the movement of water on land as well as below the surface of the earth [16].

aquifer's ability to diffuse water through the formation affects its yield [12, 14], a term referred to as permeability. Thus, according to Sophocleous [15], an aquifer's permeability measures the connectivity of spaces within the aquifer, which allows for water to move through the aquifer. Aquifers can be superimposed by layers of soil, or other layers of rocks, and they may also sit on similar or different layers of rocks. These top and bottom layers may be permeable (allowing water to flow through the aquifer), thus known as aquitards or impermeable [allowing no movement of water through (into or out of) the aquifer] and, therefore, known as aquicludes (**Figure 1**).

Aquifers are, therefore, classified into three categories (unconfined, semi-confined and confined) based on their level of porosity and permeability. Unconfined aquifers are those aquifers where there is usually a hydraulic connection between surface and groundwater resources, or where there is connection between groundwater and water in the vadose (unsaturated) zone. A semi-confined aquifer (aquifer 2 in **Figure 1**) is an aquifer superimposed by a semi-permeable aquitard (thus allowing limited movement of water between the aquifer and the above environment) and underlain by an impermeable aquiclude, thus restricting the movement of water from below. A confined aquifer, often referred to as a "*fossil aquifer*," is an aquifer that is completely closed-off, covered by an aquiclude on all the sides of the aquifer, thus completely closing off the water that saturated the aquifer during its formation [15]. As mentioned earlier, aquifers and their associated groundwater resources offer a vital relief in water-scarce areas, and similar to surface waters, these resources also cross political borders.

1.2 Transboundary aquifers

Transboundary aquifers (TBAs) are aquifers that are located beneath the surface of more than one country and are, thus, shared by those countries. Transboundary groundwater resources, therefore, sustain ecosystem services and ecological functions

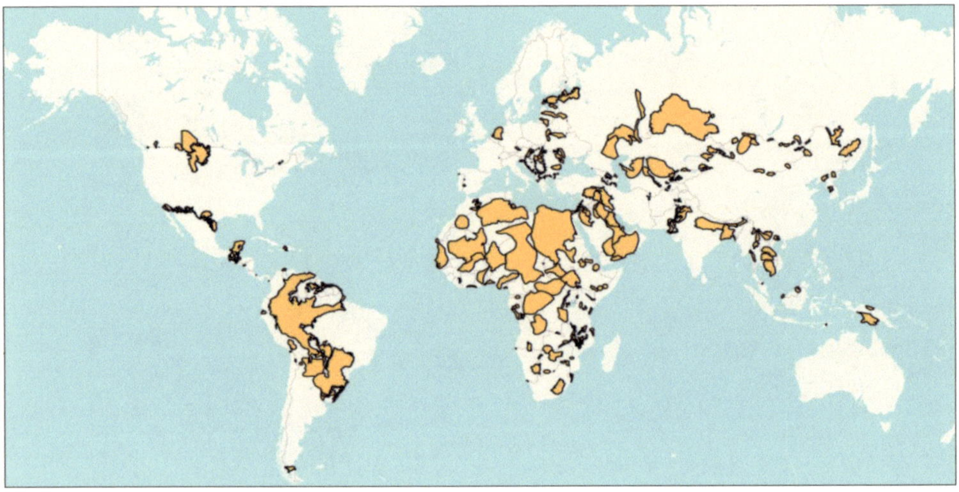

Figure 2.
Transboundary aquifers of the world [17].

of ecosystems reliant on groundwater in more than one country. Similar to national aquifers, transboundary aquifers are vulnerable to over abstraction and contamination; however, their exposure to pollution and risk of over abstraction increases with each border that the resource crosses. According to the International Groundwater Resources Assessment Centre (IGRAC), approximately 468 transboundary aquifer systems have been identified worldwide, as depicted in **Figure 2**. Of these, 135 transboundary aquifers were identified in the Americas, 130 in Asia, 106 in Africa, and 97 in Europe [17].

Of the 106 transboundary aquifers found in Africa, 24 aquifers are shared by the 12 states representing the Southern African Development Community (SADC): Angola, Botswana, Democratic Republic of Congo, Eswatini, Lesotho, Malawi, Mozambique, Namibia, Tanzania, South Africa, Zambia and Zimbabwe (**Figure 3**).

Of the 24 transboundary aquifers found within the SADC region, nine are shared by South Africa and its neighbouring states. This study, however, focuses only on two of the biggest transboundary aquifers shared by South Africa and its neighbouring countries. The study focuses on the Karoo Sedimentary Transboundary Aquifer (KSTA) shared by South Africa and Lesotho, as well as the Stampriet Transboundary Aquifer System (STAS), shared by Botswana, Namibia and South Africa (AF1 and AF5 in **Figure 3**).

1.3 Remote sensing (RS) techniques and geographical information system (GIS) applications in groundwater studies

The use of earth observation applications has recently gained momentum [18–23]. The techniques offer unique opportunities to simultaneously study large surface areas, making it possible to extrapolate and correlate isolated spatial data. Moreover, earth observation techniques can provide large-scale (regional or even global) information that would otherwise be impossible to collate in a short space of time. Thus, these techniques make it possible to study variations in groundwater storage (GWS) in areas where ground-based data are not available, as well as for inaccessible areas. The availability of earth observation applications coupled with the advent and popularity of geographic information systems (GIS) has made investigations in

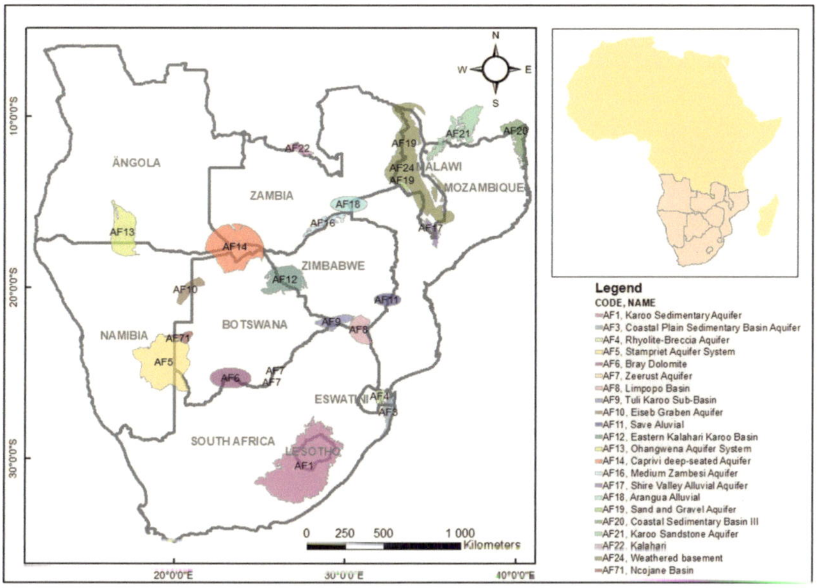

Figure 3.
Map of the 24 transboundary aquifers located in the SADC region.

groundwater studies a possibility in largely arid and semi/hyper-arid locations where the skills for ground-based observations are not available [24–27].

Spatio-temporal remote sensing provides global near-real-time, automated, cost-effective data that are accessible and accurate [24, 28–31]. The aim of this study was to explore the use of remote sensing (RS) techniques, coupled with the application of geographical information systems (GIS) to assess the spatio-temporal variability of groundwater resources of the Karoo Sedimentary Transboundary Aquifer as well as the Stampriet Transboundary Aquifer System.

2. Materials and methods

Southern Africa is a semi-arid region characterised by low and erratic rainfall, occasional flooding, high evaporation rates, cold winters and prolonged droughts [2]. The region is mostly underlain by the 2.5 million km^2 Kalahari Basin [32] and the 500,000 km^2 Karoo Basin [33]. The Karoo Basin covers the entirety of the Kingdom of Lesotho, and most of South Africa, whereas the Kalahari Basin is located in parts of Angola, most of Botswana, small patches in the Democratic Republic of Congo, some parts of Eastern Namibia and parts of the Northern-Cape in South Africa, as well parts of Zambia and small patches in Zimbabwe [34]. The KSTA is located within the Karoo Basin, and the STAS is located within the Kalahari Basin.

2.1 The Karoo Sedimentary Transboundary Aquifer

The Karoo Sedimentary Transboundary Aquifer (KSTA) is 165,936 km^2, covering all of Lesotho, most of the Eastern Cape, and parts of Kwa-Zulu Natal, the Free State and the Northern Cape provinces in South Africa (**Figure 4**). It is the largest TBA between South Africa and a neighbouring state, with an estimated population of just approximately 2.2 million people [35].

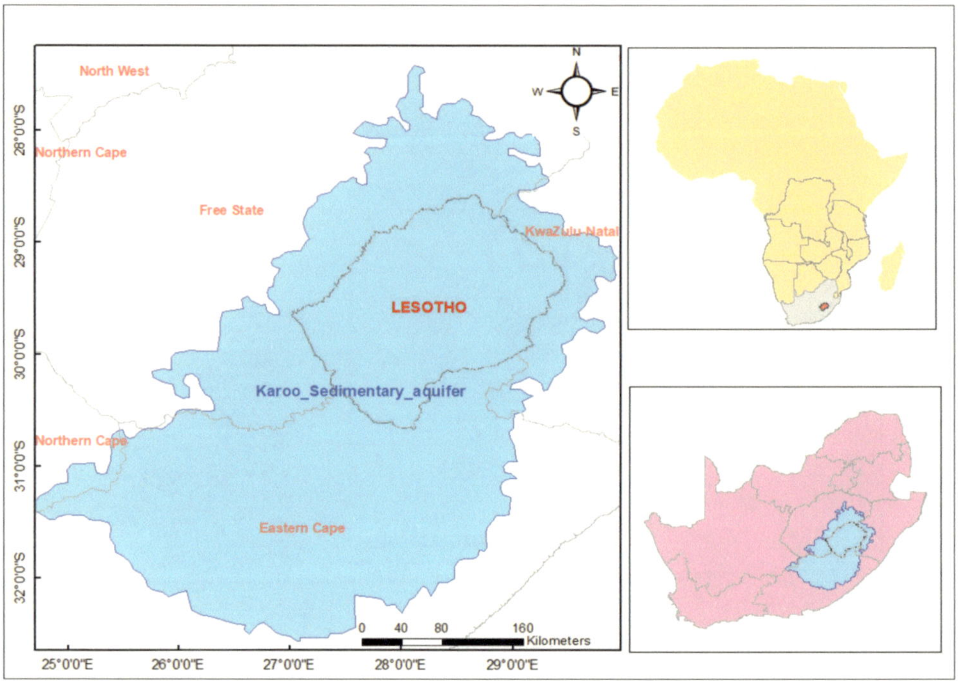

Figure 4.
Map of the Karoo-Sedimentary Transboundary Aquifer (KSTA) showing the location of the aquifer.

2.2 The Stampriet Transboundary Aquifer System

The Stampriet Transboundary Aquifer System (STAS) is 102,401 km^2 and straddles the shared border between Botswana, Namibia and South Africa. Most of the aquifer is in central Namibia, which accounts for 73% of the surface area and just over 90% of the population (total population estimated at 50,000) [36]. Botswana makes up 19%, while a small part of the Northern Cape in South Africa (which is entirely a National Park) constitutes the rest (**Figure 5**). The aquifer is named after the village of Stampriet, which is located in Hardap, Namibia.

The two transboundary aquifers are located below the Orange-Senqu transboundary river basin shared by Botswana, Lesotho, Namibia and South Africa (**Figure 6**). The majority of the Karoo Sedimentary TBA lies beneath the Orange-Senqu River Basin, while the entire Stampriet TBA system is located beneath the transboundary river basin, with some of the upper aquifers hydraulically linked to the basin, whose management, conservation, governance, protection and development are administered through the Orange-Senqu River Commission (ORASECOM).

2.3 Groundwater storage

The study used National Aeronautics and Space Administration (NASA)'s Goddard Earth Sciences Data and Information Services Center (GES DISC) to access and download data used in the study. NASA's Goddard Earth Sciences Data and Information Services Center developed the Goddard Interactive Online Visualisation and analysis Infrastructure or "Giovanni." Giovanni is an application that allows the visualisation of selected geophysical parameters [37]. It supports

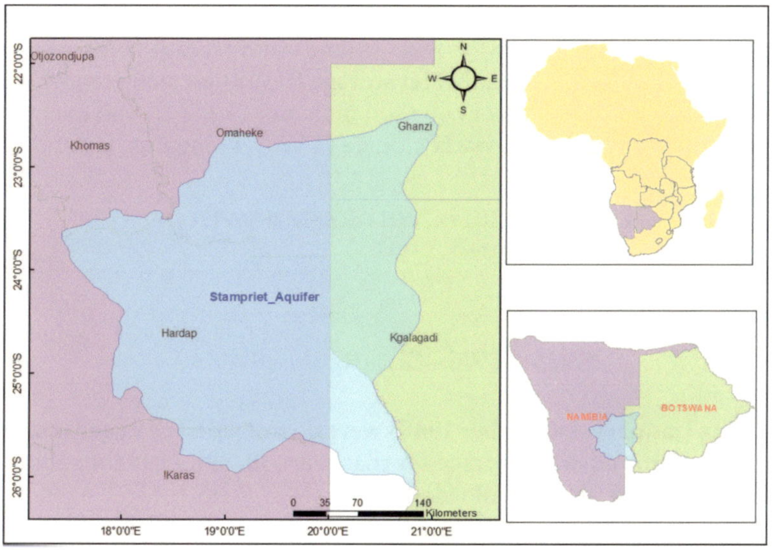

Figure 5.
Map of the Stampriet Transboundary Aquifer System (STAS) showing the location of the aquifer.

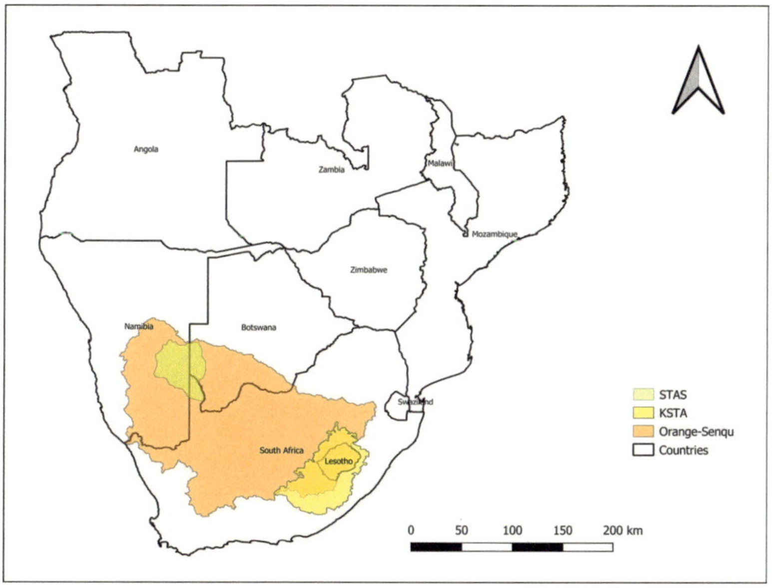

Figure 6.
Map of the Orange-Senqu basin showing the location of the KSTA and the STAS within the basin.

single- and multi-parameter visualisations as well as statistical analysis, providing an interactive interface for comparing data from a multitude of sources [37]. The data used in the study were accessed and downloaded from Giovanni. Groundwater storage data were retrieved from the Global Land Data Assimilation System Version 2 (GLDAS-2), which currently covers data from January 1948 to March 2022. The study utilised data from January 1948 to December 2020, covering a span of 72 years.

The study utilised gridded GLDAS Model CLMS025 groundwater storage data maps of 0.25° resolution (27. 2 km) to map groundwater storage in the two selected aquifers in Southern Africa. Groundwater storage (GWS) is calculated by subtracting snow water equivalent (SWE), root zone soil moisture (RZSM), and canopy interception storage (CIS), from the total water storage (TWS), using Eqs. (1) and (2) below.

$$TWS = CIS + GWS + RZSM + SWE \tag{1}$$

Thus,

$$GWS = TWS - CIS - RZSM - SWE. \tag{2}$$

Annual data (January–December 1948) were downloaded as raster image files, showing the total groundwater storage for that year [38, 39]. Data for subsequent years (1949–2020) were also downloaded.

A geographical information systems' application was used for the extraction and manipulation of the data. Before the extraction of data, shapefiles for each aquifer were created, and these were saved as vector files. For each of the study area aquifers, a shapefile was used for masking/clipping the GLDAS-gridded raster images. For each aquifer, groundwater storage data were extracted in 10-year intervals for a period of seven decades (1948–2018), a total of 72 years when including the years 2019 and 2020.

The raster images for two periods to be compared (e.g., 1948 and 1958) were differenced using the map algebra function in the GIS application. This involved subtracting the raster 1948 from the 1958 raster. The differenced map represents the change in storage for that period. The maps showing the change in groundwater storage for each aquifer and the surrounds were saved as .PNG files, and the images are presented in Section 3, showing areas of significant recharge and discharge in each study area. The storage data are provided in the raster file as depth in millimetres. The depth was converted to volume in cubic kilometres using the following Eq. (3) below:

$$\text{Volume}\left(\text{km}^3\right) = \frac{\text{depth}\,(mm)}{1{,}000{,}000} \times \text{Area of the aquifer}\left(km^2\right). \tag{3}$$

1 km = 1,000,000 mm; hence, in (2), there is division by 1,000,000 in order to convert the millimetres to kilometres.

Since each raster image grid cell size in this case is 0.25° × 0.25°, which is equivalent to 27. 2 km × 27.2 km, the size of each grid is, therefore, an area of 739.84 km^2. This area was multiplied by the number of grid cells for each study area in order to calculate the size of each aquifer.

For the purposes of this study, groundwater is defined as subsurface freshwater found within confined and unconfined aquifers. Thus, changes in groundwater depth refer to changes in the depth of the water table, whereas changes in groundwater volume refer to changes in the volume of groundwater within the demarcated area of each transboundary aquifer. Moreover, mentioning groundwater volume or storage capacity of the Karoo (KSTA) or the Stampriet (STAS) refers to the volume of groundwater within each aquifer/aquifer system.

3. Results and discussion

The insufficient availability of groundwater data (such as storage, quality and recharge and discharge patterns) results in a knowledge gap regarding the variability of groundwater resources globally. This is exacerbated by overexploitation as well as well as contamination that is usually unaccounted for and that can also go unnoticed for decades.

Thus, the limited information on the quantity (availability) and quality (contamination) of groundwater resources, coupled with anthropogenic activities as well as climate change, threatens the protection and management as well as the sustainable use of groundwater resources [40]. Accordingly, the use of earth observation or remote sensing techniques in monitoring groundwater resources has become invaluable in groundwater monitoring studies.

3.1 The Karoo-Sedimentary Transboundary Aquifer (KSTA)

Groundwater storage within the area of the Karoo Sedimentary aquifer is depicted in **Figure** 7. The volume of groundwater is presented over a period of 72 years, from the 1 January, 1948 to 31 December, 2020.

The volume of groundwater in the area fluctuated (between 95 and 120 km^3) throughout the study period, showing periods of decreasing and increasing volumes, with 1975 and 1978 depicting higher volumes of 128 and 134 km^3, respectively. The aquifer's largest increase in volume was recorded between 1970 and 1980, and again between 2000 and 2010, with the largest decrease in volume recorded between 2010 and 2020. The monthly fluctuations in groundwater volume are depicted in **Figure 8**.

Groundwater volumes were higher between January and July where average volumes were above 105 km^3. There was a steady decline in average volumes between August and November each year, and volumes start to increase again in December each year. These seasonal fluctuations in groundwater volumes correspond to the rainy season in the region. The region receives summer rainfall, starting in December through to March/April each year.

3.1.1 Spatial changes in groundwater resources of the KSTA

The KSTA is the largest transboundary aquifer that South Africa shares with a neighbouring state. Thus, changes in groundwater storage will not be uniform throughout the entire surface area of the aquifer. During a given period, some areas of the aquifer may experience increased recharges, whereas other areas of the same aquifer experience a discharge. **Figure 9** shows the spatial variation in groundwater storage depth of the KSTA.

Figure 9 shows how the depth of groundwater in the aquifer changed over time. Areas that increased in depth as well as areas where depth was reduced are depicted across the aquifer. The figure shows that large parts of the aquifer experienced a decline in groundwater storage depth between 1978 and 2007. The figure also shows that recharge and discharge is not uniform throughout the surface of the aquifer. Moreover, even though there was some form of discharge of groundwater from the aquifer between 1948 and 1967, at that time discharge occurred in small areas. Furthermore, between 1948 and 1957, large discharge of groundwater occurred in parts of the Free State, the Eastern Cape, as well as small areas around the border of the Northern Cape Province (refer to **Figure 4** showing the location of the aquifer), as revealed by the negative changes in groundwater depth.

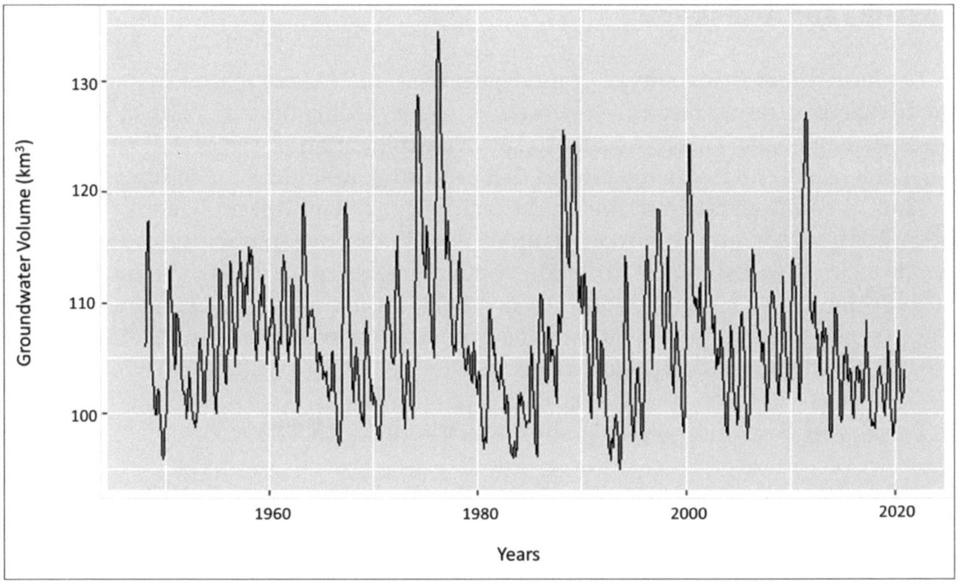

Figure 7.
Average groundwater volume in the KSTA, 1948–2020.

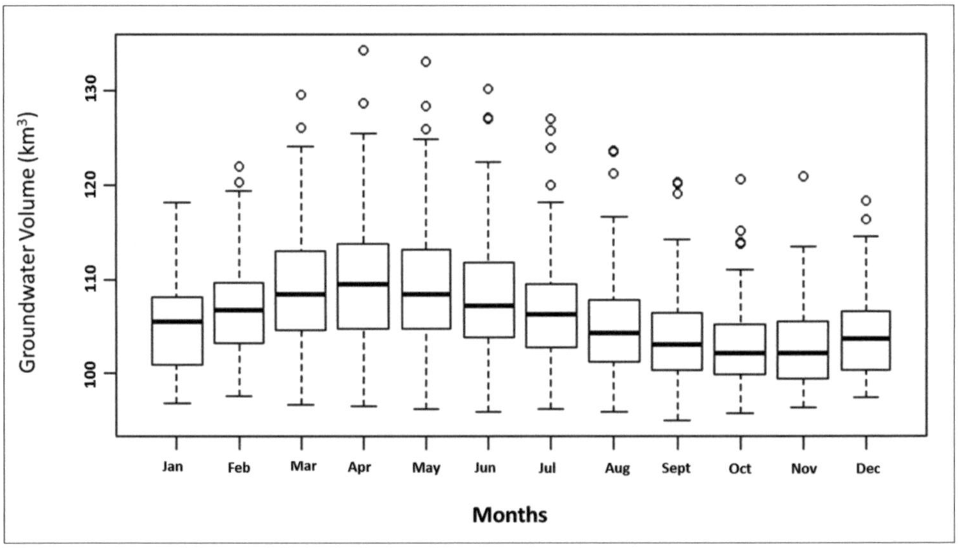

Figure 8.
Groundwater volume in the KSTA, 1948–2020.

In the same decade, positive changes in depth were observed throughout the entire surface of the aquifer; however, increased recharge occurred largely in Lesotho. In the next decade, it can be seen from **Figure 9** that areas of discharge covered the north-eastern parts of the aquifer and became more pronounced in Lesotho, spreading to neighbouring parts of the Eastern Cape as well as Kwa-Zulu Natal, whereas recharge was observed throughout the remaining area of the aquifer. This changed again between 1968 and 1977, with the decade experiencing the

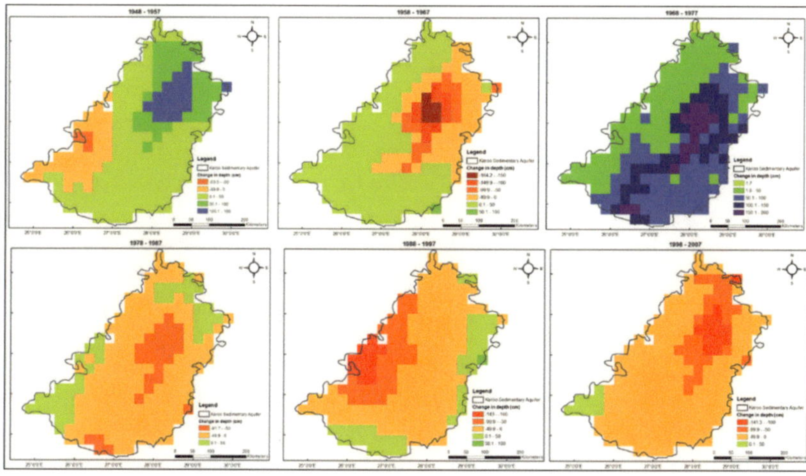

Figure 9.
Spatial changes in groundwater storage of the Karoo Sedimentary TBA.

highest recharge (largest positive changes in groundwater depth) recorded in the study. Although the decade (1968–1977) was dominated by higher recharge and low to almost non-existent discharge, this was short-lived. The rate of discharge (shown by the decreasing depth of water in the aquifer) increased and eventually became higher than the rate of recharge (which is represented by increasing depth) in the subsequent years from 1978 to 2007. Areas of recharge currently account for less than 10% of the total surface area of the Karoo transboundary aquifer. The graph also shows that the largest discharge of groundwater from the Karoo aquifer occurs in Lesotho. Similar conclusions were reached in previous studies in California [41, 42], China [43, 44], India [45–50], Iran [51], the Indus Basin transboundary aquifer [52], Kenya [53] and the North-Western Sahara Aquifer System (NWSAS) [54], where groundwater storage was found to be declining over time.

3.2 The Stampriet Transboundary Aquifer System (STAS)

Groundwater storage within the area of the Stampriet aquifer is depicted in **Figure 10**. The volume of groundwater in the area is presented over a period of 72 years, from 1 January, 1948 to 31 December, 2020.

The volume of groundwater fluctuated between 30 and 37 km^3, with some years (1950–1951, 1954–1955,1957–1958, 2000–2001, and 2005–2006) depicting volumes above 37 km^3. The volume of groundwater in the aquifer fluctuated throughout the study period, showing periods of declines and increases in volume. The aquifer's largest increase in volume was recorded between 1970 and 1980, and the largest decrease was recorded between 1980 and 1990. Moreover, the largest volumes of groundwater were recorded between 1975 and 1978, where values were above 45 and 47 km^3, respectively. The monthly fluctuations in groundwater volume are depicted in **Figure 11**.

Groundwater volumes were higher between January and July where average volumes are 33 km^3 or above, with a slight decline in average volumes recorded between August and December each year.

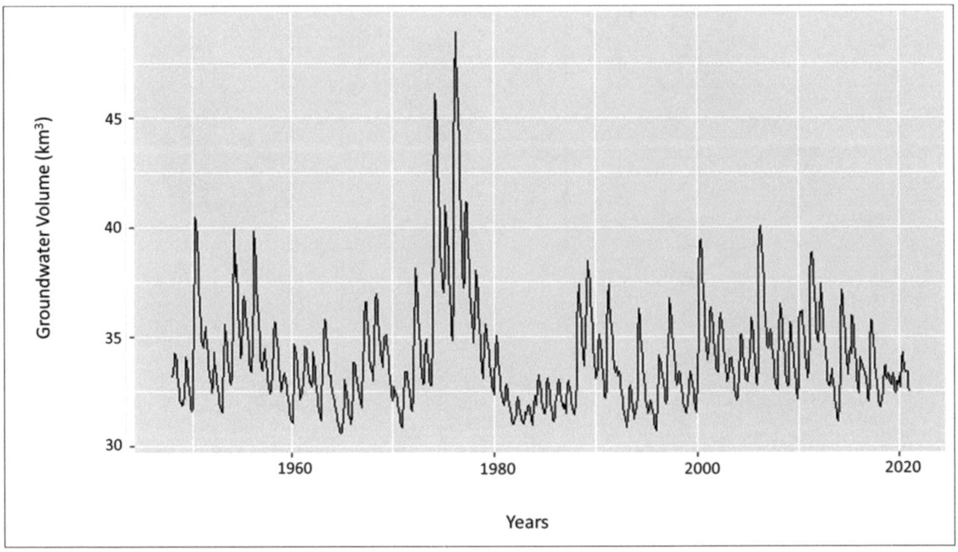

Figure 10.
Average groundwater volume in the STAS, 1948–2020.

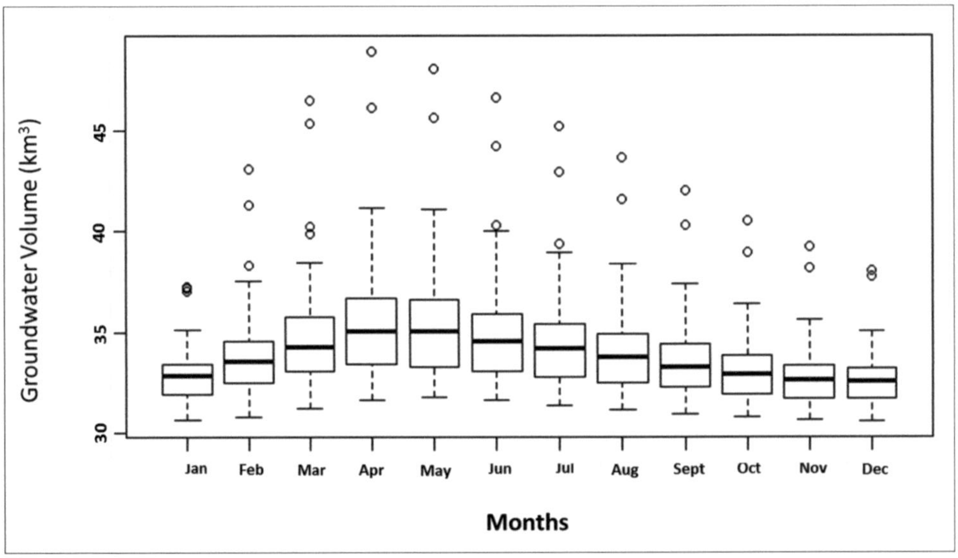

Figure 11.
Groundwater volume in the STAS, 1948–2020.

3.2.1 Spatial changes in groundwater resources of the STAS

Similar to the KSTA, the STAS is also large; thus, changes in groundwater storage will not be uniform throughout the entire surface area of the aquifer. The spatial changes that occurred in the Stampriet (STAS) are depicted in **Figure 12**, showing how the depth of groundwater in the aquifer changed over time.

Areas that increased in depth as well as areas where depth was reduced are depicted across the aquifer. Although **Figure 12** shows that large parts (approximately

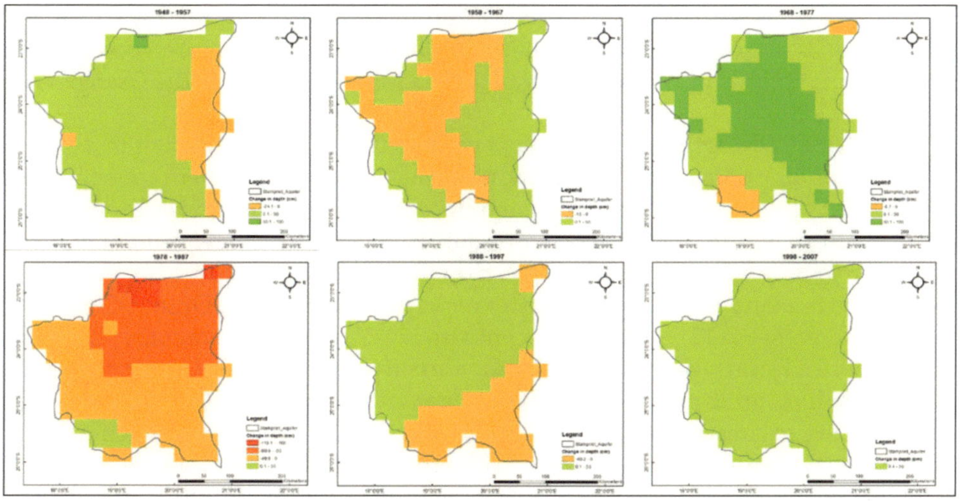

Figure 12.
Spatial changes in the groundwater storage of the Stampriet TBA.

90%) of the aquifer experienced a decline in storage depth between 1978 and 1987, areas of recharge increased between 1988 and 2007. During the first decade of study (1948–1957), the Stampriet was dominated by areas of recharge (0.1–50 cm increase in depth), with a small portion of the aquifer showing points of discharge (−24.1 to 0.1 cm decrease in depth) nested along the border of Botswana and South Africa (**Figure 12**). In the next decade (1958–1967), areas of discharge increased and included parts of the aquifer underlying Namibia. The largest positive change in storage depth (increase in depth) occurred between 1968 and 1977, with large areas of the aquifer experiencing increased depth, and only a small part of the aquifer experiencing a minimal decrease in depth. The greatest loss in storage in the Stampriet were recorded between 1978 and 1987, where only a small portion of the aquifer recorded an increase in depth, whereas the remaining parts of the aquifer experienced a loss in depth. Some areas of the aquifer recorded losses of 50 cm and above, with some areas also recording losses above 100 cm. A period of recovery was documented between 1988 and 1997 in Namibia, where the majority of the aquifer's surface area depicted an increase in depth, whereas decreasing depths were observed in parts of the aquifer underlying Botswana, South Africa as well as some parts of southwestern Namibia. Although there were periods when the Stampriet recorded large reduction in depth (1978–1987), the overall positive change in depths reflects increased recharge. The two aquifers mentioned above have experienced fluctuations in groundwater depth and volume over the 72-year study period. Both the aquifers experienced an increase in depth between 1968 and 1977, whereas the largest losses were recorded in both aquifers between 1978 and 1987.

3.3 Conservation, development and management of the KSTA and the STAS

The governments of the Republic of Botswana, the Kingdom of Lesotho, the Republic of Namibia and the Republic of South Africa established the Orange-Senqu River Commission Agreement (ORASECOM agreement) on 3 November, 2000 [55]. The agreement was founded on the Helsinki Rules [56], the UN Watercourses

Convention [57] as well as the revised SADC protocol on shared water resources [58], with the aim of serving as technical advisor to the parties on matters relating to the development, utilisation and conservation of the water resources of the Orange-Senqu Basin [55].

The ORASECOM agreement lists cooperation as the first obligation of the parties, as stipulated in article 7(1) of the agreement. The obligation to cooperate is an important provision, as it makes it possible for member states to jointly develop, conserve and govern their shared water resources in order to achieve mutual benefit. In addition to the obligation to cooperate, the ORASECOM agreement imposes on member states a duty to protect and preserve the ecosystem of the Orange-Senqu watercourse, article 7(12) of the agreement also imposes a duty on member states to individually and jointly take all the relevant measures to protect and preserve the river system from its sources and headwaters to its common terminus, and article 5(2)(6) imposes a duty to control and reduce pollution. Article 7(2) of the agreement further imposes a duty on watercourse states to utilise the resources of the transboundary catchment in an equitable and reasonable manner in line with the 2000 revised SADC protocol, with a view to attaining an optimal and sustainable use of the resource, in order to ensure its protection.

Member states are further instructed to take all suitable actions to avoid the causing of significant transboundary harm to any other party in line with the revised SADC protocol. The ORASECOM agreement further imposes on member states a duty to include an environmental impact assessment (EIA) study as part of their due diligence and to also address the possible effects of any planned activity on the social, cultural, economic as well as the natural environment. Reflecting the 1997 water-courses convention, as well as the 2000 revised SADC protocol, the ORASECOM agreement also stipulates in articles (7(5-11, and 16) the rules and procedures to be followed by member states in planning and implementing activities (within their territories), which might have adverse effects on the shared watercourse or on other member states.

The ORASECOM agreement established the commission on the Orange-Senqu Basin. Through cooperation, the commission is able to protect and manage their shared water resources in an equitable and sustainable manner with the consideration of each member state's interest in the shared resource. In addition to the legal status of the commission stipulated in the first article of the agreement, the agreement also makes provision for member states to establish river commissions with regard to any part of the Orange-Senqu (article 1(4) and further states that such agreements will be subordinate to the ORASECOM agreement. Furthermore, the agreement makes provision for the establishment of joint commissions and committees on the protection, conservation, development and governance of their respective watercourses. As a result, the ORASECOM established the Groundwater Hydrology Committee (GWHC), which is responsible for all the groundwater-related matters of the Orange-Senqu Catchment (ORASECOM, 2021), thus making provision for the protection and governance of shared groundwater resources through the joint committees of cooperation.

In May 2017, ORASECOM's Groundwater Hydrology Committee (GWHC) as well as the Technical Task Team (TTT) supported the proposal to establish within the ORASECOM structure a multi-country cooperation mechanism (MCCM) for the governance and management of the Stampriet Transboundary Aquifer System [59]. The MCCM is meant to facilitate cooperation among states in the development, monitoring, governance and management of the STAS [59, 60]. The cooperation mechanism was established as part of the STAS-Governance of Groundwater Resources in Transboundary Aquifers (GGRETA) Project with the aim of the mechanism becoming a joint mechanism advising member states on the management of the STAS [59].

The results of the study show that there is great spatio-temporal variability in the groundwater resources within the two transboundary aquifers shared by South Africa and its neighbouring countries. Moreover, the seasonal variability in groundwater resources in the two transboundary aquifers is in line with findings from previous studies that found that groundwater variability is largely influenced by climatic variations [61–66], predominantly precipitation [67, 68], as well as evapotranspiration [69, 70]. Moreover, as shown in previous studies [18, 45, 61, 71–81], data acquired through remote sensing techniques can be exceptionally applicable in hydrogeological studies where *in situ* data are insufficient. Furthermore, in addition to the ORASECOM Groundwater Hydrology Committee, the cooperation mechanism mentioned above is the first governance mechanism to ever be incorporated in a river basin organisation, making it possible for the commission to manage the water resources of the Orange-Senqu conjunctively, following Integrated Water Resources Management (IWRM) processes.

Since only six international agreements and arrangements have been concluded[1] on the utilisation, sound management and governance of over 400 transboundary aquifers globally, compared to over 3600 agreements and arrangements that have been signed on the 279 shared surface waters [82], the law regulating transboundary groundwater resources is, thus, gravely lagging and, as such, still in its initial stages. Therefore, through the establishment of joint commissions such as the multi-country cooperation mechanism responsible for the development and management of the STAS, the equitable use and sustainable management of transboundary groundwater resources is possible not only in South Africa, but globally too.

The study has shown that incorporating the management of groundwater resources through the establishment of groundwater specific committees within existing surface water agreements can ensure the protection, as well as manage equitable and sustainable utilisation of shared groundwater resources.

4. Conclusions

Monitoring groundwater resources using remote sensing techniques alleviates the pressure on the routine monitoring of sparsely distributed monitoring wells. Moreover, the use of satellites allows for the acquisition of data at spatio-temporal scales that would otherwise be inaccessible.

Although the historical data used in the study do not take into consideration anthropogenic impacts on changes in groundwater storage, the results show that groundwater resources in the region are declining, and these changes can be attributed to the effects of climate change.

Thus, with improved monitoring (taking into consideration anthropogenic activities) coupled with sound management, the protection, conservation and sustainability of groundwater resources in Southern Africa may be improved.

The study shows that the poor management of water resources (local, regional and international) as a result of the lack of data and information as well as the lack of resources (human, financial and infrastructural) can be overcome with the use of

[1] Agreement on the Al-Sag/Al-Disi Aquifer, Franco-Swiss Genevese Aquifer, Iullemeden Aquifer System, Guarani Aquifer System, North-Western Sahara Aquifer System, and the Nubian Sandstone Aquifer.

remote sensing techniques coupled with the application of geographical information systems. The results obtained in this study are applicable at all spheres of water management and governance (freshwater resources including both surface and groundwater resources, as well as brackish and marine water resources), from the local, municipal, provincial, national as well as regional and international scales. Moreover, similar results can be achieved in other fields of study too wherever RS data are available and accessible and where GIS is applicable.

Acknowledgements

The author would like to thank NASA's Goddard Earth Sciences Data and Information Services Center for making the data available and accessible, augmenting data where it would otherwise not be available and thus enabling research.

Conflict of interest

The author declares no conflict of interest.

Author details

Malebo Matlala
University of South Africa, Florida Science Campus, South Africa

*Address all correspondence to: matlamd1@unisa.ac.za

References

[1] Matlala MD, Taylor JC, Harding WR. Development of a Diatom Index for Wetland Health, WRC Report No KV 270/112011. pp. 1-186. Available from: https://www.wrc.org.za/wp-content/uploads/mdocs/KV%20270-11.pdf

[2] Davies BR, Day JA. Vanishing Waters. South Africa: University of Cape Town Press; 1998

[3] Loáiciga HA. Sustainable groundwater exploitation. International Geology Review. 2002;**44**:12. DOI: 10.2747/0020-6814.44.12.1115

[4] Stephan RM. Trans-boundary water resources.in Arab environment water: Sustainable management of a scarce resource. In: El-Ashry M, Saab N, Zeitoon B, editors. Report of the Arab Forum for Environment and Development. 2010. Available from: https www.databank.com.lb/docs/Water-Sustainable%20Management%20of%20a

[5] Matlala MD. Geo-hydrology of Transboundary Aquifers and International Water Law. In: Proceedings of the XVI World Water Congress, International Water Resources Association IWRA, 29 May – 03 June 2017; Cancun, Mexico. Available from: www.iwra.org/member/congress/resource/ABSID503_Revised_Geohydrology_of_Transboundary_Confined_Aquifers_and_International_Water_Law.pdf

[6] Christelis G, Hunger G, Mulele O, Mannathoko I, van Wyk E, Braune E, et al. Towards Transboundary Aquifer Management in Southern Africa. 2010. Available from: www.siagua.org/sites/default/files/documentos/documentos/sur_africa.pdf

[7] Kahil T, Albiac J, Fischer G, Strokal M, Tramberend S, Greve P, et al. A nexus modeling framework for assessing water scarcity solutions. Current Opinion in Environmental Sustainability. 2019;**40**:72-80. DOI: 10.1016/j.cosust.2019.09.009

[8] Uddin M, Nash S, Olbert IA. A review of water quality index models and their use for assessing surface water quality. Ecological Indicators. 2021;**122**. DOI: 10.1016/j.ecolind.2020.107218

[9] Foster SSD, Chilton PJ. Groundwater: The processes and global significance of aquifer degradation. Philosophical Transactions of the Royal Society of London. 2003;**B 358**:1957-1972. Available from: www.ncbi.nlm.nih.gov/pmc/articles/PMC1693287/pdf/14728791.pdf

[10] Margat J. Hydrogeological Mapping, in Groundwater Studies, An international guide for hydrogeological investigations. In: Kovalevsky VS, Kruseman GP, Rushton KR, editors. Ihp-Vi, Series on Groundwater. Vol. 3. 2004. p. 319

[11] Braune E, Xu Y. Groundwater management issues in southern Africa-An IWRM perspective. Water SA, IWRM Special Edition. 2008;**34**:6. DOI: 10.4314/wsa.v34i6.183672

[12] Conti IK, Gupta J. Global governance principles for the sustainable development of groundwater resources. International Environmental Agreements. 2016;**16**:849-871. DOI: 10.1007/s10784-015-9316-3

[13] Peralta RC. Assuring a Long-Term Groundwater Supply: Issues, Goals and Tools, USU Cooperative Extension Service Factsheet EL-289. Logan, UT: Utah State University; 1995. 6 p, 289. Available from: https://www.researchgate.net/

publication/237254562_Assuring_a_Long_Term_Groundwater_Supply_Issues_Goals_and_Tools

[14] Bouwer H. Artificial recharge of groundwater: Hydrogeology and engineering. Hydrogeology Journal. 2002;**10**:121. DOI: 10.1007/s10040-001-0182-4

[15] Sophocleous M. Interactions between groundwater and surface water: The state of the science. Hydrogeology Journal. 2002;**10**:52-67. DOI: 10.1007/s10040-001-0170-8

[16] Morris BL, Lawrence ARL, Chilton PJC, Adams B, Calow RC, Klinck BA. Groundwater and its Susceptibility to Degradation: A Global Assessment of the Problem and Options for Management. Early Warning and Assessment Report Series, RS. 03-3 United Nations Environment Programme, Nairobi, Kenya. Available from: https://wedocs.unep.org/bitstream/handle/20.500.11822/8035/Groundwater_INC_cover.pdf?sequence=3&amp%3BisAllowed

[17] IGRAC (International Groundwater Resources Assessment Centre). Transboundary Aquifers of the World, Update 2021, Scale 1:50 000 000, Special edition for the 2021 ISARM Conference. TBAMap_2021. Available from: https://www.un-igrac.org/transboundary-aquifers-world-map

[18] Becker MW. Potential for satellite remote sensing of groundwater. Ground Water. 2006;**44**:306-318. DOI: 10.1111/j.1745-6584.2005.00123.x

[19] Brunner P, Franssen H-JH, Kgotlhang L, Bauer-Gottwein P, Kinzelbach W. How can remote sensing contribute in groundwater modeling? Hydrogeology Journal. 2007;**15**:5-18. DOI: 10.1007/s10040-006-0127-z

[20] Iqbal N, Hossain F, Lee H, Akhter G. Satellite gravimetric estimation of groundwater storage variations over Indus Basin in Pakistan. IEEE Journal of Selected Topics in Applied Earth Observations and Remote Sensing. 2016;**9**:3524-3534. DOI: 10.1109/JSTARS.2016.2574378

[21] Katpatal YB, Rishma C, Singh CK. Sensitivity of the gravity recovery and climate experiment (GRACE) to the complexity of aquifer systems for monitoring of groundwater. Hydrogeology Journal. 2018;**26**:933-943. DOI: 10.1007/s10040-017-1686-x

[22] Rahaman MM, Thakur B, Kalra A, Ahmad S. Modeling of GRACE-derived groundwater information in the Colorado River basin. Hydrology. 2019;**6**(1):19. DOI: 10.3390/hydrology6010019

[23] Thakur JK, Singh SK, Ekanthalu VS. Integrating remote sensing, geographic information systems and global positioning system techniques with hydrological modelling. Applied Water Science. 2016;**7**:1595-1608. DOI: 10.1007/s13201-016-0384-5

[24] Abdelkareem M, El-Baz F, Askalany M, Akawy A, Ghoneim E. Groundwater prospect map of Egypt's Qena Valley using data fusion. International Journal of Image and Data Fusion. 2012;**3**:169-189. DOI: 10.1080/19479832.2011.569510

[25] Gaber A, Abdelkareem M, Abdelsadek IS, Koch M, El-Baz F. Using InSAR coherence for investigating the interplay of fluvial and aeolian features in arid lands: Implications for groundwater potential in Egypt. Remote Sensing. 2018;**10**(6):832. DOI: 10.3390/rs10060832

[26] Abdalla F, Moubark K, Abdelkareem M. Groundwater potential

mapping using GIS, linear weighted combination techniques and geochemical processes identification, west of the Qena area, Upper Egypt. Journal of Taibah University for Science. 2020;**14**:1350-1362. DOI: 10.1080/16583655.2020.1822646

[27] Odeh T, Mohammad AH, Hussein H, Ismail M, Almomani T. Over-pumping of groundwater, in Irbid, governorate, northern Jordan: A conceptual model to analyze the effects of urbanization and agricultural activities on groundwater levels and salinity. Environment and Earth Science. 2019;**78**:40. DOI: 10.1007/s12665-018-8031-0

[28] Benjmel K, Amraoui F, Boutaleb S, Ouchchen M, Tahiri A, Touab A. Mapping of groundwater potential zones in crystalline terrain using remote sensing, GIS techniques, and multicriteria data analysis (case of the Ighrem region, Western anti-atlas, Morocco). Water. 2020;**12**:471. DOI: 10.3390/w12020471

[29] Hussein H, Conker A, Grand M. Small is beautiful but not trendy: Understanding the allure of big hydraulic works in the Euphrates-Tigris and Nile waterscapes. Mediterranean Politics. 2020;**27**(3):297-320. DOI: 10.1080/13629395.2020.1799167

[30] Abdelkareem M, El-Baz F. Analyses of optical images and radar data reveal structural features and predict groundwater accumulations in the central Eastern Desert of Egypt. Arabian Journal of Geosciences. 2015;**8**:2653-2666. DOI: 10.1007/s12517-014-1434-7

[31] Abrams W, Ghoneim E, Shew R, LaMaskin T, Al-Bloushi K, Hussein S, et al. Delineation of groundwater potential (GWP) in the northern United Arab Emirates and Oman using geospatial technologies in conjunction with Simple Additive Weight (SAW), Analytical Hierarchy Process (AHP), and Probabilistic Frequency Ratio (PFR) techniques. Journal of Arid Environments. 2018;**157**:77-96. DOI: 10.1016/j.jaridenv.2018.05.005

[32] Summerfield MA, Thomas DSG, Shaw PA. The Kalahari Environment. Cambridge: Cambridge University Press; 1991. p. 298. ISBN 0 521 37080 9

[33] Adams S, Titus R, Pieterson K, Tredoux G, Harris C. Hydrochemical characteristics of aquifers near Sutherland in the Western Karoo, South Africa. Journal of Hydrology. 2001;**241**:91-103. DOI: 10.1016/S0022-1694(00)00370-X

[34] Talma AS, Verhagen BT, Tredoux G. Kalahari groundwater isotope synthesis: In the use of isotope hydrology to characterize and assess water resources in South(ern) Africa. In: Abiye T, editor. WRC Report No TT 570/13. 2013. pp. 111-122. Available from: https://www.wrc.org.za/mdocs-posts/the-use-of-isotope-hydrology-to-characterize-and-assess-water-resources-in-southern-africa

[35] United Nations, Department of Economic and Social Affairs, Population Division. World Population Prospects 2019, Online Edition. Rev. 1. 2019. Available from: https://population.un.org/wpp/

[36] United Nations Educational, Scientific and Cultural Organization (UNESCO). Stampriet Transboundary Aquifer System Assessment. Governance of Groundwater Resources in Transboundary Aquifers (GGRETA) - Phase 1 Technical Report. 2016. Available from: http://groundwaterportal.net

[37] Giovanni User Guide. 2022. Available from: https://giovanni.gsfc.nasa.gov/giovanni/doc/UsersManualworkingdocument.docx.html

[38] Li B, Beaudoing H, Rodell M. NASA/GSFC/HSL. GLDAS Catchment Land

Surface Model L4 daily 0.25 x 0.25 degree V2.0. Greenbelt, Maryland, USA: Goddard Earth Sciences Data and Information Services Center (GES DISC); 2018. DOI: 10.5067/LYHA9088MFWQ

[39] Li B, Rodell M, Sheffield J, Wood E, Sutanudjaja E. Long-term, non-anthropogenic groundwater storage changes simulated by three global-scale hydrological models. Scientific Reports. 2019;**9**:10746. DOI: 10.1038/s41598-019-47219-z

[40] MacDonald A, Bonsor H, Dochartaigh B, Taylor R. Quantitative maps of groundwater resources in Africa. Environmental Research Letters. 2012;**7**(2):024009. DOI: 10.1088/1748-9326/7/2/024009

[41] Famiglietti JS, Lo M, Ho SL, Bethune J, Anderson KJ, Syed TH, et al. Satellites measure recent rates of groundwater depletion in California's Central Valley. Geophysical Research Letters. 2011;**38**:L03403. DOI: 10.1029/2010GL046442

[42] Scanlon BR, Longuevergne L, Long D, D. Ground referencing GRACE satellite estimates of groundwater storage changes in the California Central Valley, USA. Water Resources Research. 2012;**48**:4. DOI: 10.1029/2011WR011312

[43] Huo A, Peng J, Chen X, Deng L, Wang G, Cheng Y. Groundwater storage and depletion trends in the loess areas of China. Environmental Earth Sciences. 2016;**75**:1167. DOI: 10.1007/s12665-016-5951-4

[44] Yin W, Hu L, Jiao JJ. Evaluation of groundwater storage variations in northern China using GRACE data. Geofluids. 2017;**2017**:1-13. Article ID 8254824. DOI: 10.1155/2017/8254824

[45] Bhanja SN, Rodell M, Bailing Li B, Saha D, Mukherjee A. Spatio-temporal variability of groundwater storage in India. Journal of Hydrology. 2017;**544**:428-437. DOI: 10.1016/j.jhydrol.2016.11.052

[46] Chanu S, Munagapati H, Tiwari VM, Kumar A, Elango L. Use of GRACE time-series data for estimating groundwater storage at small scale. Journal of Earth System Science. 2020;**129**:215. DOI: 10.1007/s12040-020-01465-2

[47] Joshi N, Rahaman M, Thakur B, Shrestha A, Kalra A, Gupta R. Assessing the effects of climate variability on groundwater in northern India. In: Ahmad S, Murray R, editors. World Environmental and Water Resources Congress 2020: Groundwater, Sustainability, Hydro-Climate/Climate Change, and Environmental Engineering. 2020. DOI: 10.1061/9780784482964.005

[48] Sarkar T, Kannaujiya S, Taloor AK, Ray PKC, Chauhan P. Integrated study of GRACE data derived interannual groundwater storage variability over water stressed Indian regions. Groundwater for Sustainable Development. 2020;**10**:100376. DOI: 10.1016/j.gsd.2020.100376

[49] Islam Z, Ranganathan M, Bagyaraj M, Singh SK, Gautam SK. Multi-decadal groundwater variability analysis using geostatistical method for groundwater sustainability. Environment, Development and Sustainability. 2022;**24**:3146-3164. DOI: 10.1007/s10668-021-01563-1

[50] Srivastava S, Dikshit O. Analysis of groundwater storage (GWS) dynamics and its temporal evolution for the Indo-Gangetic plain using GRACE data. Remote Sensing Applications: Society and Environment. 2022;**25**:100685. DOI: 10.1016/j.rsase.2021.100685

[51] Semiromi MT, Koch M. Analysis of spatio-temporal variability of surface–groundwater interactions in the Gharehsoo river basin, Iran, using a coupled SWAT-MODFLOW model. Environmental Earth Sciences. 2019;**78**:201. DOI: 10.1007/s12665-019-8206-3

[52] Ali S, Wanga Q, Liu D, Fu Q, Rahaman M, Faiz MA, et al. Estimation of spatio-temporal groundwater storage variations in the lower transboundary Indus Basin using GRACE satellite. Journal of Hydrology. 2022;**605**:127315. DOI: 10.1016/j.jhydrol.2021.127315

[53] Ouma YO, Aballa DO, Marinda DO, Tateishi R, Hahn M. Use of GRACE time-variable data and GLDAS-LSM for estimating groundwater storage variability at small basin scales: A case study of the Nzoia River Basin. International Journal of Remote Sensing. 2015;**36**(22):5707-5736. DOI: 10.1080/01431161.2015.1104743

[54] Ahmed M, Sultan M, Wahr J, Yan E. The use of GRACE data to monitor natural and anthropogenic induced variations in water availability across Africa. Earth-Science Reviews. 2014;**136**:289-300. DOI: 10.1016/j.earscirev.2014.05.009

[55] Agreement Between the Governments of the Republic of Botswana, the Kingdom of Lesotho, the Republic of Namibia, and the Republic of South Africa on the Establishment of the Orange-Senqu River Commission (ORASECOM) 03 November 2000

[56] Helsinki Rules on the Uses of the Waters of International Rivers, Adopted by the International Law Association (ILA) at the 52nd Conference, Helsinki, Finland. Aug 1966

[57] Convention on the Law of Non-Navigational Uses of International Watercourses, May 21, 1997, 36 ILM 700

[58] Southern African Development Community (SADC). Revised Protocol on Shared Watercourses in the Southern African Development Community. 2000

[59] Operationalization of The Stampriet Transboundary Aquifer System (STAS) Multi-Country Cooperation Mechanism (MCCM). Terms of reference. 2018

[60] Nijstena G, Christelis G, Villholth KG, Braune E, Gaye CB. Transboundary aquifers of Africa: Review of the current state of knowledge and progress towards sustainable development and management. Journal of Hydrology: Regional Studies. 2018;**20**:21-34

[61] Henry CM, Allen DM, Huang J. Groundwater storage variability and annual recharge using well-hydrograph and GRACE satellite data. Hydrogeology Journal. 2011;**19**:741-755. DOI: 10.1007/s10040-011-0724-3

[62] Hu K, Awange JL, Khandu, Forootan E, Goncalves RM, Fleming K. Hydrogeological characterisation of groundwater over Brazil using remotely sensed and model products. Science of the Total Environment. 2017;**599-600**:372-386. DOI: 10.1016/j.scitotenv.2017.04.188

[63] Frappart F, Papa F, Güntner A, Tomasella J, Pfeffer J, Ramilliena G, et al. The spatio-temporal variability of groundwater storage in the Amazon River basin. Advances in Water Resources. 2019;**124**:41-52. DOI: 10.1016/j.advwatres.2018.12.005

[64] Huang Z, Yeh PJ-F, Pana Y, Jiao JJ, Gong H, Li X, et al. Detection of large-scale groundwater storage variability over the karstic regions in Southwest China. Journal of Hydrology. 2019;**569**:409-422. DOI: 10.1016/j.jhydrol.2018.11.071

[65] Thomas BF, Famiglietti JS. Identifying climate-induced groundwater

depletion in GRACE observations. Scientific Reports. 2019;**9**:4124. DOI: 10.1038/s41598-019-40155-y

[66] Hua KX, Awange JL, Kuhna M, Nanteza J. Inference of the spatio-temporal variability and storage potential of groundwater in data-deficient regions through groundwater models and inversion of impact factors on groundwater, as exemplified by the Lake Victoria Basin. Science of the Total Environment. 2021;**800**:49355. DOI: 10.1016/j.scitotenv.2021.149355

[67] Rust W, Holman I, Corstanje C, Bloomfield J, Cuthbert M. A conceptual model for climatic teleconnection signal control on groundwater variability in the UK and Europe. Earth-Science Reviews. 2017;**177**:164-174. DOI: 10.1016/j.earscirev.2017.09.017

[68] Joshi N, Kalra A, Lamb KW. Land–ocean–atmosphere influences on groundwater variability in the South Atlantic–Gulf Region. Hydrology. 2020;**7**:71. DOI: 10.3390/hydrology7040071

[69] Chen P, Arnold PG, Srinivasan R, Volk M, Allen PM. Surveying ground water level using remote sensing: An example over the Seco and Hondo Creek watershed in Texas. Groundwater Monitoring and Remediation. 2006;**26**(2):94-102. DOI: 10.1111/j.1745-6592.2006.00077.x

[70] Li B, Rodell M, Famiglietti JS. Groundwater variability across temporal and spatial scales in the central and north-eastern U.S. Journal of Hydrology. 2015;**525**:769-780. DOI: 10.1016/j.jhydrol.2015.04.033

[71] Rodell M, Chen J, Kato H, Famiglietti JS, Nigro J, Wilson CR. Estimating groundwater storage changes in the Mississippi River basin (USA) using GRACE. Hydrogeology Journal. 2007;**15**:159-166. DOI: 10.1007/s10040-006-0103-7

[72] Shamsudduha M, Taylor RG, Longuevergne L. Monitoring groundwater storage changes in the highly seasonal humid tropics: Validation of GRACE measurements in the Bengal Basin. Water Resources Research. 2012;**48**:2. DOI: 10.1029/2011WR010993

[73] Castellazzi P, Martel R, Galloway DL, Longuevergne L, Rivera A. Assessing groundwater depletion and dynamics using GRACE and InSAR: Potential and limitations. Groundwater. 2016;**54**(6):768-780. DOI: 10.1111/gwat.12453

[74] Nanteza J, de Linage CR, Thomas BF, Famiglietti JS. Monitoring groundwater storage changes in complex basement aquifers: An evaluation of the GRACE satellites over East Africa. Water Resources Research. 2016;**52**(12):9542-9564. DOI: 10.1002/2016WR018846

[75] Feng W, Shum CK, Zhong M, Pan Y. Groundwater storage changes in China from satellite gravity: An overview. Remote Sensing. 2018;**10**:674. DOI: 10.3390/rs10050674

[76] Sheffield J, Wood EF, Pan M, Beck H, Coccia G, Serrat-Capdevila A, et al. Satellite remote sensing for water resources management: Potential for supporting sustainable development in data-poor regions. Water Resources Research. 2018;**54**(12):9724-9758. DOI: 10.1029/2017WR022437

[77] Meghwal R, Shah D, Mishra V. On the changes in groundwater storage variability in Western India using GRACE and Well observations. Remote Sensing in Earth Systems Sciences. 2019;**2**:260-272. DOI: 10.1007/s41976-019-00026-6

[78] Resende TC, Longuevergne L, Gurdak JJ, Leblanc M, Favreau G, Ansems N, et al. Assessment of the impacts of climate variability on total water storage across Africa: Implications for groundwater resources management. Hydrogeology Journal. 2019;**27**:493-512. DOI: 10.1007/s10040-018-1864-5

[79] Neves MC, Nunes LM, Monteiro JP. Evaluation of GRACE data for water resource management in Iberia: A case study of groundwater storage monitoring in the Algarve region. Journal of Hydrology: Regional Studies. 2020;**32**:100734. DOI: 10.1016/j.ejrh.2020.100734

[80] Rateb A, Scanlon BR, Pool DR, Sun A, Zhang Z, Chen J, et al. Comparison of Groundwater Storage Changes from GRACE Satellites with Monitoring and Modeling of Major U.S. Aquifers. Water Resources Research. 2020;**56**:12. DOI: 10.1029/2020WR027556

[81] Kim KH, Liu Z, Rodell M, Beaudoing H, Massoud E, Kitchens J, et al. An evaluation of remotely sensed and In situ data sufficiency for SGMA-scale groundwater studies in the Central Valley, California. Journal of the American Water Resources Association. 2021;**57**(5):664-674. DOI: 10.1111/1752-1688.12898

[82] Eckstein G, Eckstein Y. A hydrogeological approach to transboundary ground water resources and international law. American University International Law Review. 2003;**19**(2):201-258. Available from: https://digitalcommons.wcl.american.edu/cgi/viewcontent.cgi?article=1163&context=auilr